The Key to the Universe

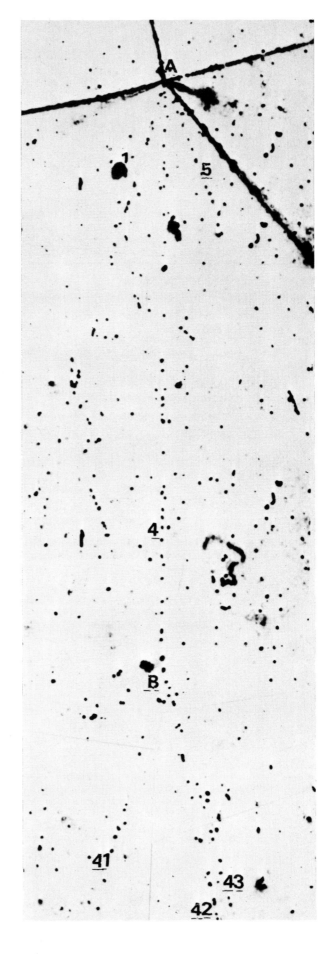

A kind of matter sought more fervently than gold appears in the first picture of the track of a probable particle with 'charm'. Travelling down the page from the point of creation at A it leaves a sketchy dotted track (4) in a piece of photographic emulsion, before it breaks up into other particles at point B. (See also page 124.)

Nigel Calder
The Key to the Universe

A REPORT ON THE NEW PHYSICS

THE VIKING PRESS   NEW YORK

Published in 1977 by The Viking Press
625 Madison Avenue, New York, N.Y. 10022

LIBRARY OF CONGRESS CATALOGING
IN PUBLICATION DATA

Calder, Nigel.
Key to the universe.
1. Physics. 2. Cosmology.
I. Title.
QC24.5.c34   539   76-51766
ISBN 0-670-41270-8

Printed in the United States of America
Set in Times Roman

Author's note

To write this book I had first to learn a lot of fresh physics, and I hope that experience of my own difficulties has helped to make the way smoother for the reader. I am grateful to my tutors – the physicists and astrophysicists in Europe and the USA who were unstinting in their help during the preparation of this book and the television programme which it accompanies. As more than 200 individuals gave direct assistance I regrettably cannot acknowledge them all by name, nor do full justice to their researches in the pages that follow. Special collective thanks are due to the directors and staff of the principal accelerator laboratories, namely the European Organization for Nuclear Research (CERN), the Fermi National Accelerator Laboratory, the Brookhaven National Laboratory and the Stanford Linear Accelerator Center. More personal thanks go to Eric Burhop, Richard Feynman, Stephen Hawking, Gerard 't Hooft, Martin Rees, Abdus Salam, Christopher Llewellyn-Smith and Steven Weinberg for their interactive help in getting the story straight. Any remaining kinks in it are, of course, entirely my fault. Finally I am grateful to the BBC and its overseas partners who backed the television production and made the travels possible.

Note on units used (see also pages 17 to 19)
One billion = 1,000 million
'Mass-energy units' are MeV (million electron-volts). One MeV is the energy acquired by an electron accelerated through a million volts. It is equivalent to an energy of 16×10^{-14} joules or a mass of 18×10^{-28} grams.
Temperatures are given in degrees absolute ($^\circ$K) which are equal to degrees centigrade, but start from 0°K $= -273^\circ$C.

The television programme, *The Key to the Universe*, was first transmitted on BBC2 on 27 January 1977. It was made by the BBC as a coproduction with BRT (Brussels), KRO (Hilversum) and WTTW (Chicago). The executive producer was Alec Nisbett, assisted by Vivienne King. David Feig was the principal film cameraman and the film editor was Christopher Woolley. Graphic design was by Alan Jeapes; visual effects by Ian Scoones; studio design by Andy Dimond and studio direction by John Gorman. The programme was written and presented by Nigel Calder.

Contents

The Key to the Universe

1 Life-maker

Human understanding of the nature of the universe and the laws of creation was advancing in the 1970s with a speed that left the experts breathless. Inhabitants of a small planet were gazing across a vast ocean of space and discerning the beginning of time. They thought they could also make out strange places called black holes, where time seemed to come to an end. Others peered with even greater rewards into the micro-universe – into the realm of particles, much smaller than atoms, from which nature made galaxies and brains. There they found deep connections between cosmic forces and the qualities of matter on which they acted.

My book describes those advances. In detail, they extended mankind's knowledge of the contents of the material universe but, behind the question 'What?', the question 'Why?' was becoming more insistent. The key to the universe might be a brief set of equations or diagrams that encompassed all the large-scale and small-scale workings of the cosmos and showed their logical connections with the character of space and time. In the mid-1970s leading theorists were speaking openly of that objective, the grand synthesis, as if it might almost be within reach.

Others declared it could never be quite attained, or gave technical reasons for thinking that the time was not ripe. But few could remain aloof from the excitement of the search for 'naked charm', for example, and the demonstration it would give that people understood the micro-universe remarkably well. The quest for the key, or for shining pieces of it at least, always led to new adventures.

By the mid-1970s, it seemed a pity that the public remained largely unaware or unappreciative of the adventures of the day. With Alec Nisbett of the BBC, I visited the big machines and observatories and consulted the leading theorists. Our travels took us to eight countries, to venerable universities, to brand-new accelerator laboratories and to remote observatories. Our report became, in part at least, a snapshot of a rewarding moment in a long struggle. Recent events demanded special attention and the confirmation of 'naked charm' occurred while I was in the middle of writing this book. Yet there were also satisfying clicks as discoveries of a hundred years past fell into place in a broader scheme of the universe, while people tried to pick the lock of ignorance.

Instrument-making in the 1970s, when nature's secrets were becoming ever costlier to lay bare. At the CERN laboratory near Geneva a mechanical mole bored a seven-kilometre tunnel through the rocks of France and Switzerland, to create a merry-go-round for sub-atomic particles.

Accelerator at Serpukhov, USSR. The picture shows a section of the 1½-kilometre ring of large electromagnets designed to hold particles in orbit during their acceleration. From 1967 to 1972 Serpukhov's machine was the most powerful in the world.

Going to extremes

Almost absent-mindedly, governments and taxpayers had equipped the searchers to find out how the universe was constructed. Great mirror-telescopes adorned unmisty mountain-tops in California and the Caucasus, Chile and New South Wales. Wide radio dishes were planted more often on the plains, while costly little satellites circled the Earth to sense rays from the universe that would not reach the ground.

Astronomy was big science but not as big as high-energy physics, devoted to the exploration of the micro-universe. Any thorough account of the universe would have to explain why nature had mass-produced particles of certain kinds, wherewith to build atoms, stars, planets and living things. Looking deeply into matter required the most elaborate instruments ever conceived and engineered for scientific purposes.

The aim was to uncover unsuspected phenomena, and to test the prevailing theories about the forces of nature to breaking point. That meant going to extremes: packing energy into very small volumes to achieve the most intense conditions possible. Therefore, dotted around the world, were machines devoted to the creation of forms of matter previously unknown, and to the ever-closer scrutiny of known forms. The machines used the electric force to accelerate particles until their energy of motion was enormous. Then the particles were allowed to crash into material targets, or into one another, while elaborate instruments – particle-counters and devices called bubble chambers – recorded the products of the impacts.

Nature itself provided high-energy particles, in the cosmic rays raining on the Earth from outer space. Investigators using this free supply made many important findings about the constituents of matter. Occasional cosmic-ray particles possessed energies far surpassing anything that could be engineered on Earth. But catching them was a slow and chancy business and by the 1970s most of the discoveries were coming from the big accelerators.

Some of the important machines accelerated electrons, the same particles of matter as those used to paint pictures on a television screen. But electrons were nimble and tended to shed energy as fast as it was added to them. The highest energies were attained by accelerating protons – the nuclei of hydrogen atoms, and the principal raw material of the universe. For a while the USSR possessed the most powerful particle-factory, at Serpukhov near Moscow. But the Russians had not accomplished very much with it before, in 1972, a gigantic accelerator built by the Americans near Chicago surpassed it.

Fermilab: the Fermi National Accelerator Laboratory at Batavia, near Chicago. Its accelerator sprawls across a large area of Illinois farming country. It accelerates protons – nuclei of hydrogen atoms – to six times the energy achieved at Serpukhov. Quite apart from the complexities of the machine and its control systems (for example, right), individual experiments carried out with the machine use elaborate and expensive equipment. The laboratory has a staff of 1300 and serves several hundred visiting experimenters at any one time. Fermilab is operated by a consortium of 51 American universities and one Canadian, under contract to the US Energy Research and Development Administration.

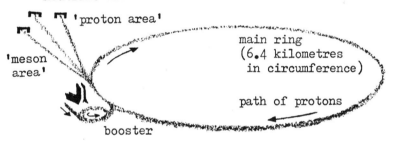

experiments

'neutrino area'

'proton area'

'meson area'

main ring
(6.4 kilometres
in circumference)

path of protons

booster

Fermilab's central laboratory building, sixteen storeys high, dominates the flat landscape, yet is small beside the big accelerator as seen on the previous page.

The big machine at the Fermi National Accelerator Laboratory endowed protons with so much energy that it made them 500 times heavier than normal. Standing in the Illinois prairie, the Fermilab accelerator was one of the engineering wonders of the world, a veritable cathedral of science. It ruled its protons with a thousand powerful electromagnets, which stood in a ring more than six kilometres in circumference. A preliminary acceleration system, itself more energetic than most of the world's accelerators, injected protons into the main ring. Thereafter the protons, travelling in a vacuum chamber, circulated 200,000 times around the merry-go-round of magnets, a distance farther than to the Moon and back. Each time around the ring, the protons felt kicks of radio energy that added a couple of units to their mass-energy.

Finally, like a stone from a sling, the pulse of accelerated protons was let fly out of the ring. It shot into the areas where a dozen or more experimental teams, including physicists from many countries, were waiting to see what they could do with all that energy. The experimental areas stretched away like railway sidings two kilometres long. The instruments that stood there were themselves elaborate and expensive.

High-energy research was rivalling even space research in its complexity and cost. In 1976 the Europeans brought a comparable machine into operation at the CERN laboratory, and the machine-builders were already sketching far more powerful accelerators. Critics sneered about 'expensive toys', and were genuinely baffled about how such facilities could be justified.

From research at energies which were 'out of this world' in the sense that ordinary matter did not share them, there was no sound reason to expect practical benefits such as novel power supplies. It was a question of finding meaning in the universe and hence in human existence; also of sustaining the right of each generation to question the ideas and assumptions of its predecessors. The director of Fermilab, Robert Wilson, was once being pressed by a senator to say whether his laboratory contributed to the security of the United States. He replied: 'It has nothing to do directly with defending our country except to help make it worth defending.'

The passionate science

The only secrets were nature's, and the discovery of them was an international enterprise – nowhere more explicitly than at CERN, the twelve-nation European Organization for Nuclear Research near Geneva. As it grew it spilled across the Swiss frontier into France, to find room for its great machines, developing the air of a university curiously mixed with an industrial complex. The street names at CERN left no doubt about the character and traditions of the enterprise: Newton, Rutherford, Einstein, Bohr. . . .

The atmosphere at CERN could best be savoured in the cafeteria, along with the Swiss cooking. At the crowded tables at lunchtime, you would hear conversations in many tongues but most often in broken English, one of the two international languages of physics. An Englishman might be forgiven for failing to recognise some of it as English, not so much because of the accents as because of the heavy jargon of machines and instruments, experiments and theories. The other international language, mathematics, would appear on the tables, scribbled on scraps of paper, to reinforce an argument.

There you might meet a visiting Nobel prizewinner sitting alongside young men whose aim was to prove him wrong. Troubleshooters from a contracting company would be found taking a break from their efforts to make a recalcitrant piece of machinery function as intended. All the skills that went into fundamental physics were represented in the little groups: machine-builders, theorists, experimenters and technical supporters of every kind. And people talked shop, because physics was their life.

After the meal the groups would disperse to a thousand different tasks. Theorists retired to a warren of offices, there to take up pen and paper and commune with the universe, trying to think how they would have organised the particles of matter, had they been God. Some engaged in high-flown mathematical speculations; others attended closely to experiments in progress, suggesting what an unexpected finding might mean, or helping in the interpretation of a harvest of measurements.

Machine operators went to the control rooms of the various accelerators, to relieve their colleagues at elaborate consoles which kept each pulse of particles under intensive care. CERN never slept: the machines ran twenty-four hours a day for weeks on end. If, because of any hitch, the supply of particles paused, dozens of experimenters clamoured for its restoration.

The experimenters were the front-line troops in CERN's war on ignorance: that was why the place existed, simply to bombard their equipment with high-energy particles. You literally took cover whenever the energetic particles began to fly, outside the radiation shields of massive concrete encasing the machines and the experiments. The foxholes were the counting-rooms, where racks of electronic equipment and computers kept track of what the detectors reported about the vast numbers of particles spraying through the apparatus whenever the accelerator delivered a pulse.

The world's principal particle accelerators (1976). Acceleration energy in MeV.

proton synchrotrons

Fermilab	400,000
CERN (SPS)	400,000
Serpukhov	76,000
Brookhaven (AGS)	33,000
CERN (PS)	28,000
Argonne (ZGS)	12,700
Tokyo (KEK)	12,000
Dubna	10,000
Rutherford (NIMROD)	8,000
Moscow	7,000
Berkeley (BEVATRON)	6,200

proton storage rings

CERN (ISR)	31,000

electron linear accelerator

Stanford (SLAC)	22,000

electron synchrotrons

Cornell	12,000
Hamburg (DESY)	7,500
Yerevan (ARUS)	6,100

electron annihilators

Stanford (SPEAR)	4,500
Novosibirsk (VEPP 3)	3,500
Hamburg (DORIS)	3,000
Orsay (DCI)	1,800
Frascati (ADONE)	1,500

Most of the experimenters were visitors from universities and institutes around Europe or further afield. They worked in large teams, to assemble their complicated arrays of particle-counters and then to man the counting-rooms around the clock. A single experiment, perhaps concerned only with some subtle point about what happened when one proton brushed against another, might be years in the planning and take months to run. It required extraordinary dedication, especially as the chances of making a remarkable discovery were always small and, even if your team was lucky, there were too many of you for much glory to rub off on the individual.

Why should grown men spend their days and nights like that? Why indeed should national treasuries pay out large sums to enable them to do so? It said something about a human quality which scarcely figured in the reckonings of economists or of politicians preoccupied with the material welfare of their citizens. Yet it was the quality which, for instance, deformed the US economy for the sake of putting men on the Moon. Whatever President Kennedy's precise motives were for wanting to impress the world with the *Apollo* project, the key factor in his reckoning was that the world *would* be impressed by a journey into the unknown. Still, perhaps it was less obvious that a fuzzy streak on a photograph, showing a distant galaxy or a sub-atomic particle with some easily forgettable name, might to its discoverer equal a new America.

The human urge to know was always a passion as strong as sex. It moved the Stone Age people to see what lay beyond the horizon, and by that wanderlust they spread all over the planet. And brains that evolved to outwit the elephant and the salmon had powers of inference and anticipation that did not cease to function when the hunters and gatherers sat around the fire and stared into the friendly flames. They speculated about the fires of the Sun and the volcano, and sought meaning in life and death. The children especially asked awkward fundamental questions about how and why things happened, and how the world was made.

As society became more elaborate, authorities embarrassed by such questions laid down official answers to them. They devised systems of education that would crush curiosity and so avoid its heretical or revolutionary consequences. Fortunately a few usually escaped this intellectual castration and preserved into adulthood their child-like sense of wonder. As a result the human species discovered many things.

For a long time the search for understanding was slow and often pathetic. Even the comparatively liberated philosophers of ancient Greece were deeply in error, in supposing that thinking about nature was more important than looking at it. The medieval astrologers and al-

Captains of high-energy physics: the directors of the principal accelerator laboratories. Here is Wolfgang Panofsky of the Stanford Linear Accelerator Center (SLAC) at Stanford University, California, where several major discoveries occurred in the 1970s. He was born in Berlin in 1910 and learned his physics at Princeton University and the California Institute of Technology. He became a professor at Stanford in 1951. He has been much involved in science policymaking for the US government, including service on President Kennedy's Science Advisory Committee.

John Adams, British machine-builder (left) and Léon Van Hove, Belgian theorist, are joint directors-general of CERN, the European Organisation for Nuclear Research based near Geneva. Adams, who is in charge of administration and operations, built both the first proton synchrotron of 1961 and the new super proton synchrotron (SPS) of 1976. He was born in 1920 and worked at Harwell before joining CERN in 1953. He was director-general of CERN 1960–61. Van Hove was born in Brussels in 1924 and in 1954 went to the Netherlands as professor of theoretical physics at Utrecht. He joined CERN in 1961 and became director of theoretical physics in 1966. He is now director-general in charge of research at CERN.

The creator and director of Fermilab, Robert Wilson, celebrates the successful trials of the exceptionally powerful accelerator in Illinois. He was born in 1914. For twenty years before his appointment as director of Fermilab he was at Cornell University where he built a series of electron accelerators at the Laboratory of Nuclear Studies.

chemists looked at nature intensively, but witchcraft and magic befuddled their thinking. Not until the seventeenth century, with the rise of the 'experimental philosophy', was the right balance struck between thinking and looking. The principle was simple enough: that thoughts would only be taken seriously if they conformed with what you saw by looking.

Nature was to be the arbiter, in observations and experiments, while a social system of meetings and publications ensured that every result of theory and experiment was subjected to the scrutiny of other experts. That was science, the most powerful engine ever conceived for the advancement of knowledge. Almost at once the modern search for the key to the universe was well under way, with Isaac Newton's theory of universal gravity, which stood up to astronomical measurements for two centuries.

Physics was always the master-science. The behaviour of matter and energy, which was its theme, underlay all action in the world. In time astronomy, chemistry, geology and even biology became extensions of physics. Moreover, its discoveries found ready application, whether in calculating the tides, creating television or releasing nuclear energy. For better or worse, physics made a noise in the world. But the abiding reason for its special status was that it posed the deepest questions to nature.

The public nevertheless formed illusions about the physicists. Unappreciative of the passion for understanding which drove them in their work more compellingly than any taskmaster, outsiders substituted the image of the mad scientist, impelled by a lust for power over nature and man. More favourably, but still quite falsely, others saw the physicist as a white-bearded old man, a creature set apart by his boundless knowledge and computer-like brain.

In reality physics was a young man's game, with the big advances often coming when people in their twenties and thirties overthrew the notions of their teachers. It always flourished best in a free environment where professors and politicians were willing to acknowledge the privileges of youth. So far from being complacent in his breadth of knowledge, the physicist was obsessed by what he did not know; and a brain with only computer-like qualities would not generate the imaginative leaps necessary for progress.

Crazy ideas and cosmic forces

Paradoxical though it might seem, to insist on arbitration by the experimental facts liberated thought rather than constraining it. You could now spin all kinds of outlandish theories about how nature worked without disseminating error or clogging the mind with nonsense. If the theories did not accord with the facts they would be automatically rejected by the corporate process of research. Moreover, the experimenters, looking ever more intently at nature, discovered weird goings-on which mere cogitation would never have guessed at, and the minds of the theorists were stretched to accommodate them. Theory and experimental discovery worked together like two hands on a double-ended saw.

The more you knew of the facts, the greater the challenge to the imagination. Two outstanding theorists of the twentieth century, Werner Heisenberg and Wolfgang Pauli, produced what they thought was an important piece of the key to the universe. In 1958 Pauli gave a lecture about it, which was attended by Niels Bohr, another great figure who forty years earlier had worked out how atoms were built. After some discussion, generally unfavourable to the theory, Bohr stood up and declared: 'We are all agreed that your theory is crazy. The question which divides us is whether it is crazy enough.'

Sufficiently crazy yet possibly correct ideas were not easy to come by, but physics generated them and led the human mind into unthought-of realms. The public, of course, was always having para-scientific speculations foisted upon it: a near-collision with the planet Venus, spacemen visiting the Earth in ancient times, a mysterious destructive influence near the island of Bermuda, and so on. But the perpetrators lacked the wit to conceive anything to compare with the physicist's simple radio, or the not-so-mysterious influence that made Italy collide with Switzerland and threw up the Alps – never mind anti-matter or the Big Bang.

Newton's crazy idea was that gravity was a cosmic force: that the same force which drove the apple towards the ground also operated in the heavens and steered the Earth itself. He was also aware of other forces at work, as shown by the challenge he threw down in 1704.

'There are therefore agents in nature able to make the particles of bodies stick together by very strong attractions. And it is the business of experimental philosophy to find them out.'

Those who came after him verified that not only gravity operated throughout the universe, but also the forces that made 'the particles of bodies stick together'. Indeed wherever they looked, from distant galaxies to the smallest particles of matter, they found underlying the conspicuous diversity of nature an extraordinary degree of law and order. *Everything* was evidently fashioned and ruled by the action of a few cosmic forces on a few basic particles of matter. The search for the key to the universe therefore came down to trying to understand those forces and particles – and the connections between them.

By 1970 a leading theory said that all the heavy matter of the universe, gathered in heavy nuclei that lay at the heart of all atoms, consisted of particles called quarks. In addition there were the well-known electrons, much lighter in weight; by their motions they filled most of the space in an atom and governed its chemical behaviour. The electrons had a few relatives, not figuring in the constitution of ordinary matter but nevertheless deemed to be basic particles. The quark family and the electron family accounted for all the durable matter in the universe.

There were said to be four cosmic forces, responsible for all action in the universe. Gravity was one, the builder of stars and guider of planets; one of the least of its duties was saving human beings from the distress of falling off the Earth. It acted on all particles without discrimination. The electric force, author of light, of lightning and of life, was another cosmic force, and the chief cement of all ordinary materials on our planet. It worked between particles endowed with an electric charge – which meant most of the basic particles.

Next came the strong nuclear force, binder of the atomic nucleus and firemaker for the Sun and the stars. It operated only on nuclear matter. The so-called weak force, the fourth in the short catalogue, was weak in strength but subtle in its influence. As the chief agent for altering the inherent qualities of matter, it played an essential part in creating diversity in the universe. Like gravity, it influenced all the basic particles of the quark and electron families.

The strong nuclear force was a good deal stronger than electricity, which in turn was far stronger than the weak force of gravity, in most circumstances. There was another important distinction. Gravity and the electric force were unlimited in range, but the strong nuclear force and the weak force operated only when particles were at very close quarters – less than a million-millionth of a centimetre apart.

A measure of progress was that, in the years after 1970, physicists were persuaded of the existence of two new cosmic forces – a fifty per cent increase in a very select company. One was a variant of the weak force, and it turned out to have unexpected implications for the behaviour of stars. The second was called the colour force and operated between quarks. It had nothing to do with real colour, except metaphorically, but it was the strongest force of all, surpassing even the strong nuclear force. Curiously enough, the new forces emerged in the course of successful efforts to unify some of the cosmic forces, so that in another sense you could say the number of forces was reduced. Other ideas were coming true, in the confirmation of the reality of the quarks and the discovery of the charmed quark, predicted by some theorists.

16

Thinking in tens

Besides crazy ideas, there were crazy numbers. The known universe stretched through a span of scales, from basic particles so small that no one had measured them, to galaxies so far away that their light took billions of years to reach the Earth. The human imagination could scarcely cope.

To say there were more than a billion billion billion atoms in the human body was scarcely more illuminating than to say 'lots'; to try to picture important events within the nucleus of an atom occurring in less than a billion-billionth of a second was equally incomprehensible to the human mind. At the other end of the available scales and numbers, the universe contained billions of galaxies like the Milky Way to which we belong, each with many billions of stars weighing perhaps a billion billion billion tons apiece. The age of the universe was estimated at about 15 billion years or half a billion billion seconds. (I use 'billion' to mean a thousand million.)

The human imagination might just about cope with a billion: the number of grains of sand that would fill a bucket; the number of millimetres between Athens and Rome; the population of China; the cost in dollars of a fully equipped aircraft carrier: all these are about a billion. But the intellect had to reach far beyond billions and billionths to comprehend the universe. To ask how many atoms there were in the observable universe, for example, was not an absurd question, even if the resulting number might look absurd.

On the back of an envelope a physicist easily jotted down the number of atoms in the universe. If he were showing how he did it he could set it out like this:

Atoms in a gram	say 10^{24}
Grams in a star	say 10^{33}
Stars in a galaxy	say 10^{11}
Galaxies in the universe	say 10^{10}
So, atoms in the universe	about 10^{78}

He would hasten to say that these were all rough and ready figures, and that he was assuming most of the matter was in the galaxies. But his answer, 10^{78}, could be translated into ten multiplied by itself 77 times, written as 1 followed by 78 zeros. All he did was to count 'tens', but tens that were multiplied together, not added. If he wanted to deal with very small quantities he could write, for example, that an atom was 10^{-78} of the mass of the universe, meaning one *divided* by ten 78 times.

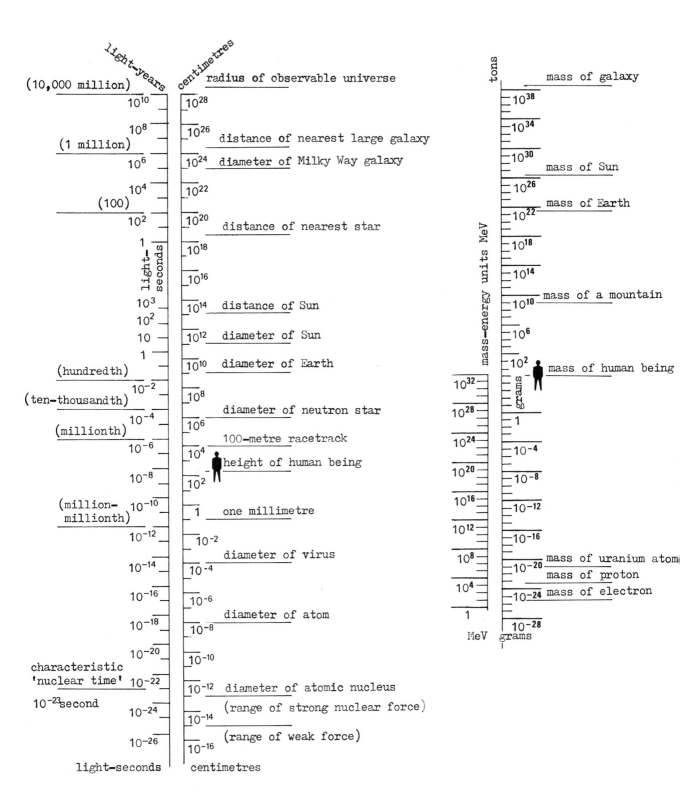

radius of observable universe
distance of nearest large galaxy
diameter of Milky Way galaxy
distance of nearest star
distance of Sun
diameter of Sun
diameter of Earth
diameter of neutron star
100-metre racetrack
height of human being
one millimetre
diameter of virus
diameter of atom
diameter of atomic nucleus
(range of strong nuclear force)
(range of weak force)

mass of galaxy
mass of Sun
mass of Earth
mass of a mountain
mass of human being
mass of uranium atom
mass of proton
mass of electron

characteristic 'nuclear time'
10^{-23} second

light-seconds   centimetres

Crazy quantities, encompassing he wide universe and the realm of very small particles and atoms, become manageable if one simply multiplies or divides repeatedly by ten, calling 1,000,000 (for example) 10^6 and $1/_{1,000,000}$ 10^{-6}. These scales are for general interest only; mastery of the quantities shown is not necessary for understanding what follows in this book. The left-hand scales show distances, in centimetres and in light-years and light-seconds, which are the distances travelled by light in the time indicated, at 30 billion centimetres per second. The right-hand scales show masses measured in tons and grams; also in mass-energy units (MeV) where one MeV or million electron-volts is equivalent to 18×10^{-28} grams, or twice the mass of an electron. The masses of particles are shown in more detail in the table below.

masses of particles in mass-energy units(MeV)	
BASIC PARTICLES	
electron family	
neutrino (electron type)	0
electron	0.5
heavy electron(muon)	105
neutrino(muon type)	0
quarks	
up	336
down	338
strange	540
charmed	1500
SOME COMPOSITES	
three quarks	
proton	938
neutron	939
lambda	1115
omega	1672
quark + anti-quark	
pion	139
K	496
'gipsy'(J/psi)	3100
* see Chapter 4	

By this comparatively simple trick, physicists roamed freely through the realms of the very large and the very small, not bothering to try to imagine the enormous numbers but dropping no stitches in the computation of the universe. And they found a wealth of useful interconnections between the quantities. For example, one large quantity was the speed of light, 300,000 kilometres per second, and it tied other quantities together. The light-year was a convenient measure of distances in the universe: about ten million million kilometres, or 13 'tens' of kilometres (10^{13}). It also had physical meaning because the light from a star 400 light-years away literally took 400 years to reach the Earth.

There was equal physical meaning when physicists considered the size and timescales of the nucleus of an atom. Events there typically took about the time that would be required for light to pass across the nucleus: 10 million-million-million-millionths (10^{-23}) of a second. There were more elaborate interconnections between cosmic quantities large and small, which brought a sense of richness and even familiarity to the physicist's reckonings.

It was his cardinal assumption that the very large and the very small were thoroughly connected. But before entering these domains, in later chapters, perhaps we should start with the middle ground, and the familiar man-sized world dominated by the electric force. In any case, a thorough understanding of the electric force, acquired over a hundred years, became the basis for more recent thoughts about the other cosmic forces.

Electricity as the life force

Learned men, like ordinary folk, were for long tempted by the idea of a kind of working version of the soul, that organised the matter in living creatures and went away at death. The processes required for keeping a plant or animal alive were awesome enough and involved organisation unknown to stars and rocks, so a great deal of polemics, as well as research, had to go into dispelling the notion that life had a force all to itself. The 'life force' was simply the electric force.

A one per cent net imbalance between positive and negative electricity in my body would shoot my head right out of the solar system. Being far stronger than gravity, the electric force would blow the Earth and the galaxies to pieces if it were not very discreet in its workings. Fortunately matter *en masse* was very strictly neutral, electrically; why it should be so was one of the deeper questions about the creation of the universe.

No great detective work was needed to establish that electricity had

James Clerk Maxwell
(1831–79), founder of
electromagnetic theory.

something to do with life. Electric shocks caused pain and twitching of the muscles, and could stop the heart, showing that the nerves and organs were sensitive to the force. Wires attached to a person's scalp revealed waves of electricity flowing across the surface of the brain. These words you are reading are being transformed into millions of electric pulses inside your head.

But all such phenomena were high-level refinements of living processes, made possible by much more basic actions of the electrical force as life-maker. It bound atoms together in special combinations and enabled them to carry out the elaborate work of sustaining life. Two great discoveries of the twentieth century consolidated this view: that all chemistry was a matter of the electrical behaviour of atoms, and that the hereditary instructions for life were embodied in a chemical code.

In the origin of life on Earth, the vital transition was from random chemistry to the systematic making of particular materials, according to schemes that could be passed from parent to offspring. The electric force united atoms in consistent ways, making molecules of predictable shape and character. But not unalterably, and the variations allowed by the electric force enabled life to evolve from primitive to higher forms. If the chemistry were too sloppy, everything would have perished of genetic mutations; too exact, and the world would have remained populated, at best, only by crude microbes. The balance between fidelity and change depended on the electric force between the atoms in living creatures being neither too weak nor too strong. The electric force could strike the right balance only in certain materials and in a certain temperature range. Life closely resembling the Earth's would be impossible on the planets Mercury and Venus (too hot) or Jupiter and beyond (too cold).

Sunlight, which created and powered all life on Earth, completed the electric force's claim to be the life-maker. The particles of light journeying from the Sun to the Earth were both products and agents of the electric force. Green plants tuned their molecules to absorb the particles of light that could penetrate the Earth's atmosphere, and put their energy to the work of living and growing.

All chemical processes, in living and non-living materials, came to be thoroughly understood in terms of the activities of electrons in atoms. The electrons, as basic particles of matter, were almost pure electric charge. The small, heavy nucleus of an atom carried electric charge of an opposite kind. Each chemical element had a fixed number of electric charges on the nucleus of each of its atoms. Accordingly, each element had a particular number of electrons in its atoms, that neutralised the charge on the nucleus. The ways in which the electrons arranged them-

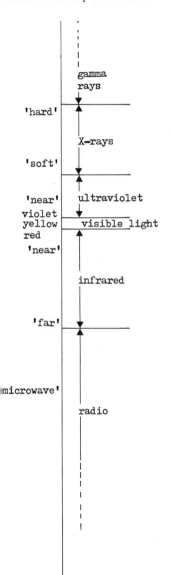

gamma rays

'hard'

X-rays

'soft'

'near'   ultraviolet

violet
yellow   visible light
red

'near'

infrared

'far'

microwave'

radio

selves in the atoms fixed the behaviour of atoms. The carbon atom, for instance, indispensable for life, had four 'loose' electrons at its exterior which could make bonds with up to four other atoms.

Metals, to give another example, proved to be made of atoms in which one or two electrons were even looser and able to detach themselves very easily, and the sea of loose electrons that pervaded a metal explained why it conducted an electric current very easily. The electrons willingly shuffled through the metal, as a moving procession of electric charges. Metals were tough for the same reason. Every atom, deprived of an electron or two, was electrically unbalanced and as they all strove to recapture their lost electrons they were drawn together like dogs after the same bone.

A first unification

The electric force became the best-understood of all the cosmic forces. Readings from spacecraft instruments many thousands of miles from the Earth, and from powerful machines which brought electric charges into the very closest confrontations that human beings could contrive, confirmed the predictions of theory with gratifying precision. The corner-stone of the theory was that two cosmic forces – electricity and magnetism – were just different aspects of a single 'electromagnetic' force, which I am calling the electric force for simplicity's sake.

Magnetism, having to do with compass needles and with picking up iron filings with permanently magnetised pieces of rock or metal, was originally quite separate from electricity – which was all about sparks and shocks and chemical batteries. How could they be the *same* force? Their unification was one of the greatest triumphs of nineteenth-century science. It began with Hans Christian Oersted's chance discovery that an electric current exerted a force on a compass needle. The story advanced, through many adventures and frustrations, to Michael Faraday's discovery that a *change* in magnetism could create an electric current – the principle of electrical generation on which modern power supplies came to depend. It culminated in the joint theory of electromagnetism, developed by James Clerk Maxwell, a 30-year-old Scotsman working in London. He threw in light for good measure.

Maxwell first confided to Faraday that light was 'electromagnetic' in 1861. The idea depended on a 'crazy' assumption that even empty space was somehow filled with electricity. But when Maxwell set out simple laws that summed up every known kind of electric and magnetic behaviour, they prescribed electromagnetic radiation travelling at the speed of light.

21

Maxwell died in 1879, ten years before his theory was gloriously vindicated by the discovery of radio by Heinrich Hertz. Here was a form of energy akin to visible light yet generated by electrical equipment. Within half a century it had given rise to radio transmissions, television, radar and radio astronomy. But the theory implied an immense rainbow of electromagnetic radiation far wider than the spectrum of visible light.

As the full wonder of Maxwell's Rainbow gradually revealed itself, the electromagnetic spectrum stretched from extremely long to very short radio waves, through the infra-red to visible light. All the colour of the world represented just one little octave of electromagnetic radiation. Beyond visible light the physicists found an energetic darkness filled with ultra-violet, X-rays and gamma-rays. They were all composed of particles of electromagnetic energy and travelled at the speed of light.

The theory contained a deeper subtlety which was not fully appreciated until the twentieth century. Unwittingly Maxwell, in his interpretation of electromagnetism, anticipated Albert Einstein's ideas of relativity. Einstein banished from the universe Isaac Newton's framework of absolute space and absolute time, declaring all motion to be 'relative'. Many pre-existing notions had to be amended, including Newton's own theory of gravity and laws of motion. But not Maxwell's theory. The 'relative' frameworks were already implicit in the interchangeability of electricity and magnetism. Indeed, the magnetism whereby the electric currents deep inside the Earth deflected a compass needle at the surface was just a version of the electric force produced by electric charges in relative motion.

Maxwell's theory was only a start. In the twentieth century three other ideas, each crazy enough in its day, combined to give an even more remarkable account of the electric force. One was Einstein's relativity; another was 'quantum mechanics' which described how sub-atomic particles behaved very differently from ordinary life-sized objects; the third was anti-matter. Before saying a little about each of them, let me jump to the conclusions that emerged, about the electric force in action.

To say that an electron, for instance, had an electric charge, simply meant that it was forever 'flashing' with invisible light. (From now on, I use the word 'light' in the broadest Maxwellian sense to cover light-like particles of all kinds.) To be more precise, for one moment in every 137 moments – a moment being a very short time – you would find a particle of light in the vicinity of the electron, and ready for action.

The flashing electron

When two electric charges influenced each other by the electric force, it was due to the fact that their attendant particles of light could skip between them. Particles of light were the carriers of the electric force, traversing the intervening space, and enabling particles not in direct contact to feel each other's presence. Galileo once remarked that the book of nature was written in a language of triangles, circles and squares. Were he alive in the twentieth century he might well have added the broken H.

Feynman diagram of the electric force

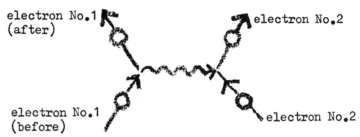

The picture is called a Feynman diagram, after Richard Feynman, an outstanding American theorist who won the Nobel Prize for showing how the electric force could be calculated with the help of diagrams like this. Here two electrons are travelling up the page and then a particle of light whizzes between them across the page. It has the effect of deflecting them: they are feeling the electric force. In practice, two electrons would swap many particles of light in the course of an encounter, like fighting ships exchanging broadsides.

Nor was this the whole story. Although it carried no electric charge itself, the particle of light had qualities in common with the electron, which allowed it to interact with it. They were implied in Maxwell's theory, but the most dramatic expression of shared qualities appeared when very energetic particles of light (gamma-rays) turned out to be capable of breaking up into two electrons with opposite electric charges – an electron and an anti-electron. But now I am running ahead of myself: we should go back and look at the evolution of ideas that made the simple picture possible.

Borrowed energy and borrowed matter

First, how did the electron come to be 'flashing'? Where did the energy come from? The central idea of quantum mechanics, as it emerged in the late 1920s, was that of 'uncertainty' developed by the German physicist Werner Heisenberg. By this principle, the 'dead' matter of the micro-universe became far livelier than anyone had imagined. As H. G. Wells put it: 'The atoms of our fathers seemed by contrast like a game of marbles abandoned in a corner of a muddy playground on a wet day.'

Werner Heisenberg (1901–76), formulator of 'uncertainty'. He was born in Germany and by the age of 23, as assistant to Max Born in Göttingen, he was working out the first thorough treatment of the mechanics of the sub-atomic realm, where objects behave in ways seemingly different from those of large-scale objects. By doing so he became one of the founders of 'quantum mechanics', a central theme of twentieth-century physics. The 'uncertainty principle', which allows sub-atomic particles to make use of undetectable energy, was for Heisenberg the epitome of quantum mechanics. In 1929 he became professor of theoretical physics at Leipzig and in 1942 director of the Kaiser Wilhelm Institute for Physics in Berlin. At that time he was a leader of the unsuccessful German efforts to develop an atomic bomb. He declared after the war that he had no wish to succeed, but his dealings with the Nazis were nevertheless ambiguous. He spent his last years in Munich, where he was director of the Max Planck Institute for Physics, 1958–71.

Heisenberg realised that particles were endowed with an amazing bonus of energy, which made them iridescent.

By the principle of uncertainty, it took a little time to know how much energy any object possessed. No test, whether by a human experimenter or by nature itself, could establish a particle's energy without allowing time for the energy to disclose itself. Conversely the uncertainty about energy became great if you asked what an object was doing during a very brief interval. And while Mother Nature blinked almost anything could happen provided it was quick enough. As a result, the economy of the micro-universe ran on 'borrowed' energy. Every financier knows that a daring man with no money can borrow some, bet it on a horse or on the changing price of copper, and hope to make a profit before he has to repay the loan. Likewise, Heisenberg found more energy available for activity in the micro-universe than you would expect by miserly accountancy.

How much 'borrowing' could nature allow, while avoiding any outward signs of misappropriation or devaluation of energy? An important quantity called Planck's Constant set the limit: the energy multiplied by the time could not exceed Planck's Constant. Thus an electron, for example, could 'borrow' a very small amount of energy to make a feeble particle of light (radio energy) that could survive for a millionth of a second or more; or a great deal of energy (gamma-ray) available only for 10^{-20} second, say.

So long as all accounts were settled promptly and in full, no laws of nature were violated. But were the particles of light created in this way real or not? Different experts had different opinions, and to play safe they called them 'virtual' particles. They were fictional to the extent that, by definition, you could never detect or measure them: if you did, it would be like demanding a settling of accounts at an inopportune moment during a financial operation. Yet the 'virtual' particles were real in so far as they could move pieces of world around – much as I might borrow a million dollars and alter the world with it before paying it back.

The full implications of uncertainty as a fact of nature became apparent only when put alongside other discoveries of the early twentieth century. One was Einstein's famous dictum, amply confirmed by later research, that mass and energy were fully interchangeable according to the formula $E = mc^2$ (E being energy, m mass and c^2 the speed of light squared). It was another great unification, denying that matter and energy were two distinct and permanent constituents of the universe. All forms of energy, whether light or motion or vibration or even the potential energy of a rock on a mountain-top before an avalanche, had a definite amount of mass associated with them. The abolition of mass could release energy. And by

24

aul Dirac, theorist of anti-
atter. He was born in
ngland in 1902 and graduated
 an electrical engineer from
ristol University at the age of
. The extraordinary notion
at for every form of matter
 equal and opposite form
uld exist, such that particle
d anti-particle would
nihilate each other
mpletely, came out of
irac's theoretical work of
28. He was awarded the
obel Prize for physics at 31,
ving meanwhile secured
ewton's former post as
ucasian professor of
athematics at Cambridge,
hich he held until 1969.
hereafter he went to Florida
 professor of physics in the
ate University in
allahassee, and continued his
eoretical researches. He has
ntributed many other
iportant ideas about the
iiverse: 'respectable' work
at fills many pages of the
iysics textbooks, and also
rther 'crazy ideas' that are
ill on the agenda, including
e prediction of magnetic
onopoles and the proposition
at force of gravity may be
owing weaker with passage
 time. Dirac's taciturn and
tiring behaviour are famous;
 his days at Cambridge, a
iit of volubility called a *dirac*
eant one word per year.

'freezing' energy you could in principle make new particles. But what kinds of particles? That was where the next crazy idea came in, the prediction of anti-matter.

Experimenters were always used to getting silly results, because of defective equipment or human error. A lot of effort went into eliminating gremlins of all kinds, before the answers could be taken seriously. So if you saw an electron swerving the wrong way under the influence of an electromagnet, you naturally assumed that you had connected the wires back to front. Unless, that is, you happened to be Carl Anderson of the California Institute of Technology. In 1932 he saw an electron turning the wrong way, among the tracks of cosmic-ray particles. He had sufficient confidence in his techniques to be sure he had discovered an electron carrying an electric charge opposite to that of an ordinary electron. It took some nerve for a 27-year-old physicist to announce such an ab-surdity. But it won him the Nobel Prize, because he had found the first piece of anti-matter.

At the time, Anderson had not read the writings of another young man, the Cambridge theorist Paul Dirac. Only a year or two before, Dirac had predicted the existence of a new kind of particle, unknown to experimental physics, with the same mass and opposite charge to an electron. 'We may call such a particle an anti-electron,' Dirac said. Years later, Heisenberg was to remark: 'I think that really the most decisive discovery in connection with the properties or the nature of elementary particles was the discovery of anti-matter by Dirac.' Heisenberg died in 1976, leaving Dirac as the sole survivor of the men who founded quantum mechanics in their youth.

In 'inventing' anti-matter Dirac was guided by his mathematics. He had imposed upon himself the very important task of reconciling two great new theories of physics, the quantum mechanics of sub-atomic behaviour, and Einstein's relativity. With self-confidence even greater than Anderson's he interpreted a minus sign in his equations as meaning the existence of negative matter rather than negative energy. It implied an extraordinary and unlooked-for symmetry at the heart of the micro-universe, such that for each particle there was an anti-particle, its opposite in every respect.

In fact anti-matter was so perfectly opposite to matter that if matching particles and anti-particles came together they would simply cancel out – annihilating one another and disappearing from the universe. They would leave behind only a puff of energy. And it was correct! Within a few decades, experimenters were confirming Dirac's crazy idea with every pulse of their big accelerators. Not just anti-electrons, but heavier anti-particles of many kinds proved to be manufacturable. People speculated about anti-stars, anti-galaxies, even an anti-universe.

But Dirac's theoretical discovery concerned deeper springs of action in the universe than the production of self-destroying curiosities. In the first place, it gave a prime law of creation. Whenever there was enough energy available, nature could create any kind of particle at any time, provided only that it created the corresponding anti-particle at the same instant. That was an excellent brief for the experimenter, who wanted to find out what the constituents of the universe might be: he could just go ahead and try to make new particles. Einstein's equation told him how much energy was needed for particles of given mass.

Nature could play the creation game unaided. The question arose whether all the matter of the universe might have been originally formed from energy in that way, in the creation of matter/anti-matter pairs. On a more accessible level, cosmic rays coming from exploding stars and colliding with particles in the Earth's atmosphere had abundant energy to create large numbers of electrons and anti-electrons, among which Anderson discovered the first anti-electron (or 'positron' as he called it).

An even more fascinating possibility arose. Nature could be creating particles and anti-particles rather casually all the time, in a hidden sort of way, using the 'borrowed' energy permitted by Heisenberg's uncertainty principle. It was a notion which implied a thorough, everyday penetration of our universe by the putative anti-universe. During the ephemeral transactions the 'borrowed' energy could be converted into any particles whatsoever, with only two restrictions. One was that the particles had to disappear like Cinderella's magic coach when midnight struck – when the period of 'borrowing' expired. The other was that matter and anti-matter had to be created equally and oppositely so that the total amount of matter in the universe did not change.

For the sake of clarity, I first described short-lived particles of light appearing in the vicinity of electrons as carriers of the electric force. But

Short-lived particles in empty space

virtual energy could exist even in empty space, because of the impossibility of measuring its lack of energy instantaneously. That energy could create short-lived pairs of particles and anti-particles. As a result even the emptiest space – say the dark emptiness between the galaxies – could be thought of as seething with short-lived particles: electrons and anti-

26

electrons, shorter-lived (because heavier) protons and anti-protons, indeed any kinds of particles known to nature.

Forces and particles

Armed with these ideas, let us look again at the picture of the electric force in action. If the particle of light passing between the two electrons were energetic enough it could itself materialise into an electron and an anti-electron. So a particle of light might be thought of as consisting of an electron and anti-electron, in which case the diagram could be drawn as follows.

Particle of light as an electron/ anti-electron 'combination'

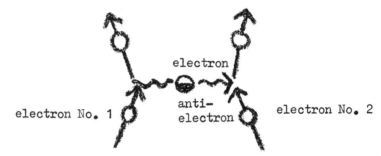

It represents the force-carrying particle of light, as if it were made from an electron and anti-electron in combination. There are disadvantages as well as advantages in this diagram. Most particles of light would have too little energy to make an electron/anti-electron pair; the implication that an electron and anti-electron might actually exist in a particle of light should not, therefore, be taken too literally. The electric charges of the electron and anti-electron cancel out and the combination works like an uncharged particle of light. If you ask, 'why don't they annihilate each other?' the answer is that they do, if you want to look at it that way, and the product of the annihilation is the particle of light.

To think of an electron and anti-electron combining to make a force-carrying particle of light emphasises that the force-carrier exists at the very interface between the universe and the anti-universe, where the creation of particles and anti-particles can occur freely. Nature keeps scrupulous accounts of how many durable particles there are in the universe – how many electrons, for instance. But it is quite casual about how many particles of light there are. You can create particles of light by striking a match any time you like. A mixture of matter and anti-matter need not, by its creation, alter the net contents of the universe; it is therefore very suitable for carrying a force whenever the need arises. Another advantage of the electron/anti-electron picture is the kinship it suggests between the force-carrier and the electrons on which it acts.

The idea of the electric force emerging from twentieth-century physics was strikingly self-contained. You could start with light and picture energetic particles of light breaking up into electrons and anti-electrons, with 'frozen energy' constituting mass. But electrons could react with light; in other words, they possessed electric charges. They could emit particles of light, too, and because one electron could feel the light emitted by another, a force operated between them – the electric force.

If the universe were made only of electrons and light physicists might claim to understand it well. But there was matter with other qualities, which could also be created from light and which was therefore also latent in light and space-time. Suppose the nature of other matter and of other cosmic forces was also to be understood in terms of particles of matter exchanging light-like particles. The light-like particles might consist of matter and anti-matter in combination and conceal within them the qualities of the particles that emitted them. Accordingly they could react with other particles that shared those qualities, thereby creating a cosmic force, in a scheme similar to the electric force.

The first great leap in this direction was taken in 1934 by a 28-year-old Japanese theorist, Hideki Yukawa, in describing the strong nuclear force. He said that short-lived force-carrying particles, heavier than electrons but lighter than protons, came into being by drawing on the 'borrowed' energy permitted by the uncertainty principle. They shuttled between the particles of the atomic nucleus, binding them together. Drawing it as a Feynman diagram, the process would look like this.

The strong nuclear force

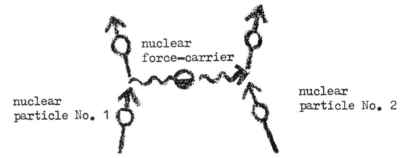

Yukawa imagined several forms of these force-carrying particles, which he called mesons. But they remained a conjecture until 1947, when the first of them was spotted by Cecil Powell of Bristol, in the cosmic rays. It soon turned out that, if you shook a proton hard enough, these force-carrying particles appeared in abundance.

In the new physics, the traditional distinction between the particles of the universe and the cosmic forces which operated upon them was disappearing. The particles and the forces came out of the same set of possibilities in nature. All of the forces could be represented by

Feynman diagrams, and all of the force-carrying particles might be thought of as consisting of mixtures of matter and anti-matter. Different cosmic forces acted upon different groups of particles, according to whether there was kinship between the qualities of the particles (such as electric charge) and the qualities implicit in the force-carrying particles themselves. Elucidating these patterns of operations in the universe, and the effort to understand why the various particles and forces existed at all, became the focus of research.

The surge forward after 1970 had to do with two salient questions about this essentially simple scheme. One was fairly subtle, and yet crucial to the understandability of the universe. If the light-like particles carrying a cosmic force concealed the qualities of the particles on which they acted should they not be able to feel the force themselves? Should the force-carriers not react with each other?

A 'yes' answer to that question would immediately make the universe wonderfully self-consistent, and confirm the kinship of forces and particles. But it was dauntingly difficult to comprehend all the possibilities of interaction implied by the 'yes' answer. Eventually the problem was cracked.

The other salient question about the scheme of particles and forces was: why was there more than one picture? Why were all particles not the same, with only one cosmic force between them? The beginnings of an answer slowly appeared. At extremely high temperatures, they might indeed all be the same, but ordinarily nature contrived to 'break the symmetry' spontaneously, creating more variety. Instead of mass-producing one particle and one kind of light, it made a small range of different particles with different qualities and different forces between them. But there remained a hidden, ideal symmetry which unified the particles and forces by making them variations of one another.

Ideas like these were not mere speculations or theoretical exercises, but flowed from a continual interplay between theory and experiment. The brightest idea of all time would have little value unless it drew on existing knowledge of the workings of the universe, and unless it made predictions that could be subjected, as Yukawa's was, to experimental tests.

ll of the forces could be presented by Feynman grams.' This picture ticipates much of what is to said later in the book. Note at it follows a convention ich makes it look unrealistic en a force causes particles to ract each other.

29

2 Starbreaker

One day in 1971 a research student, Gerard 't Hooft, walked into his professor's office in Utrecht and showed him how to do a calculation that had foxed the world's cleverest theorists. From that moment the trajectory of thought about the cosmic forces changed perceptibly, because the calculation concerned a possible link between the forces. A few years earlier an American, Steven Weinberg, and a Pakistani, Abdus Salam, had asserted that the so-called weak force was essentially akin to the electric force.

If they were right it would be the first such connection for a hundred years – the first since Maxwell unified electricity and magnetism. But there were doubts about whether their theory made any sense mathematically, until the young Dutchman found the way to solve it. Once that was accomplished, everyone was keener to check the theory.

At the CERN laboratory in Switzerland, in the same year, French engineers were completing the installation of a great new instrument named Gargamelle. It was to stand exposed to the most curious of nature's particles, the neutrinos. In half a dozen other laboratories, French, Belgian, Italian, German and British physicists were preparing to use Gargamelle for an extremely difficult experiment. They were to look for a new kind of cosmic force, predicted by the much-doubted theory.

In astronomical institutes around the world, theorists were puzzling over one of the most dramatic events in nature – the explosion of a star. They could understand the enormous release of energy in a big star in its death-throes, and the rich variety of nuclear processes to which it gave rise. But there was difficulty in explaining why the star broke so violently, blasting most of its substance into space. Something was missing from the calculations.

This chapter tells of cosmic connections that appeared between the explosions of stars, the nature of light and forces that looked, on the face of it, completely different. It also illustrates the international character of modern research and the relationship between theory and experiment; and how the progress of knowledge depends on techniques, whether theoretical techniques like Gerard 't Hooft's, or experimental techniques like those surrounding the Gargamelle instrument.

Photographs of the tracks of sub-atomic particles, recorded in the bubble chamber Gargamelle at CERN, Geneva. In this picture, three different views taken of a single event can be seen projected for measurement. Millions of these photographs have to be analysed, and the work is shared between laboratories. The measuring table shown here is at CERN.

The new cosmic force turned up in Gargamelle in 1973. To some astrophysicists it seemed to explain the bursting of an exploding star, which is why I call it the Starbreaker. The story revolved around the so-called weak force. To savour it properly one should know how the weak force contributed to the evolution of the elements and thus to our existence. In a sense human flesh is made of stardust.

The cosmic cast-list for this chapter includes, apart from the electric force and the weak force, the following basic particles.

 proton: the nucleus of the hydrogen atom and the prime constituent of heavy matter in the universe.

 neutron: very like the proton but having no electric charge; an essential ingredient of all atoms heavier than hydrogen.

two quarks, 'up' and 'down': protons and neutrons turned out to be made from them.

electron: the lightweight constituent of all atoms, with an electric charge opposite to the proton's.

heavy electron ('muon'): a particle very like an electron but weighing, for completely obscure reasons, 200 times as much.

neutrinos: particles with an amazing capacity to pass through ordinary matter, closely related to the electron and heavy electron.

 sundry force-carrying particles.

Making the elements

Every atom in the human body, excluding only the primordial hydrogen atoms, was fashioned in stars that formed, grew old and exploded most violently before the Sun and the Earth came into being. The explosions scattered the heavier elements as a fine dust through space. By the time it made the Sun, the primordial gas of the Milky Way was sufficiently enriched with heavier elements for rocky planets like the Earth to form. And from the rocks atoms escaped for eventual incorporation in living things: carbon, nitrogen, oxygen, phosphorus and sulphur for all living tissue; calcium for bones and teeth; sodium and potassium, indispensable for the workings of nerves and brains; the iron colouring blood red ... and so on.

No other conclusion of modern research testified more clearly to mankind's intimate connections with the universe at large and with the cosmic forces at work among the stars. An American nuclear physicist, William Fowler, and three British astronomers, Geoffrey Burbidge, Margaret Burbidge, and Fred Hoyle, carried out a classic study (1957–64) on how the stars made the elements. One motive for it was a wish to show that the elements had not been made in the Big Bang, at the birth of the universe. Fred Hoyle in particular was a spirited opponent of the Big Bang theory, as one of the authors of the rival Steady State theory. While Steady State's main assertion of an unchanging universe perished, the particular argument that the stars made all but the lightest elements prevailed.

A medium-sized star like the Sun was known to burn steadily in the nuclear fashion for billions of years. When it eventually began to run out of hydrogen fuel it would swell and puff away some of its contents into the surrounding space, before collapsing into a white dwarf star. Stars substantially bigger than the Sun burned much more fiercely and quickly: they were 'blue-hot' instead of white-hot. Because of their greater mass the force of gravity, acting like a pressure cooker, kept a big star hot and dense and so allowed more thorough stewing of the material of the stars.

And the big stars eventually exploded. In our galaxy, the Milky Way, such events were clearly seen only five times in a thousand years. But remains of stars that had exploded were quite plentiful. Arc-shaped clouds of dispersing debris glowed faintly among the other stars. More strident were the pulsars, the immensely compressed cores of exploding stars. They stood flashing like police beacons, each marking the scene of a cosmic accident.

Stellar explosions did remarkable things to the nuclei of atoms. The

medieval alchemists had tried to change one chemical element into another, especially hoping to make gold. Their successors in the twentieth century could say why their efforts were in vain. The essential character of an element was fixed by the number of protons (positively charged particles) in the nucleus of each of its atoms. You could transmute an element only by reaching into the nucleus itself, which the alchemists had no means of doing. But stars were playing the alchemist all the time.

Stars in a normal state, whether big or small, burned the lightest element, hydrogen, and formed from it helium, the next heaviest element. The process gave off copious energy. In very massive stars, or in less massive stars going through a phase of internal collapse, the temperature might climb high enough for the helium to burn. It changed into carbon and oxygen, with a further release of energy. Then the carbon and oxygen could burn, too, to form still heavier elements.

The escalation through the table of elements became progressively more difficult. The heavier the element, the more protons it had in each nucleus, and the more powerful was the electric repulsion between two nuclei, preventing them from fusing together. By the time you wanted oxygen to burn to make sulphur and silicon, or silicon to burn to make iron, you needed temperatures of billions of degrees so that the nuclei were colliding with sufficient frenzy to crash through the electric barrier. Iron-making marked the limit to nuclear burning in stars, and there was known to be a great deal of iron about. The Earth inherited a huge core of molten iron and meteorites often contained iron, too, all of it forged in stars. If the nuclear forces had their way, the whole universe would consist of iron.

After iron, the making of heavier elements in stars began to consume energy rather than releasing it. No star could earn a steady living that way. But in the explosion of a big star some of the enormous energy released went into building up dozens of chemical elements heavier than iron: gold, lead ... all the way through the table of elements to uranium and beyond. Even so, heavy elements remained far less abundant in the cosmos than the lighter elements.

Many of the atoms so formed, and later incorporated into the Earth, were radioactive. Their nuclei were overcharged with energy and unable to survive indefinitely. But 'not indefinitely' could mean billions of years. From uranium, thorium, potassium and other radioactive elements, energy stored during the explosions of the ancestral stars slowly trickled out into the rocks of the Earth. It generated the heat that fired volcanoes, shifted the continents and built mountains. The great creakings called earthquakes, which accompanied these processes, were thus direct con-

35

sequences – albeit greatly delayed and translocated – of those stellar explosions that made the stuff of the Earth available.

Alchemy by the weak force

The origin of the chemical elements, made from the primordial hydrogen and helium, depended on the strong nuclear force binding more and more pieces together, to make heavier and heavier atomic nuclei. But that cosmic force would be impotent but for the collaboration of another – the weak force. In those operations in stars, from the burning of hydrogen to the elaboration of gold and uranium, the weak force was busy getting rid of the electric charges from about half of the nuclear material. If that did not happen the electric force would veto all attempts to build anything more complicated than hydrogen. I call the weak force the Cosmic Alchemist; and any old-time alchemist could have told you that transmutation required a special kind of magic.

In one indispensable process the weak force altered the principal raw material of the universe, the electrically charged protons constituting the nuclei of hydrogen atoms. It could change them into electrically neutral relatives of the protons, the neutrons.

Changing the character of matter

In the burning of hydrogen in stars to make helium, four protons were united, but two of them had to change into neutrons. That was because the helium nucleus consisted of two protons and two neutrons.

Hydrogen burning: strong force binds, weak force adjusts

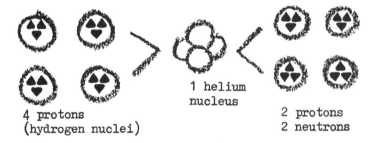

The building of the heavy elements beyond iron was done largely by neutrons, generated in large numbers in the exploding stars. They tacked themselves on to nuclei, thereby increasing the weight. But then a reverse process came into play, with some neutrons changing back into protons.

That increased the electric charge in a nucleus, so assigning it to a new chemical element. The ejection of electrons in this process constituted an important form of radioactivity, engineered by the weak force.

Building heavy elements

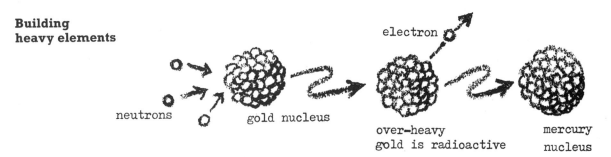

In the 1960s it became apparent that protons and neutrons were not basic particles of matter, but consisted of quarks. The next chapter will have much more to say about the quark theory and how it developed. But protons and neutrons were said to be built from two types of quarks called, ungracefully, *up* and *down*. A proton consisted of two up quarks and one down quark, while a neutron had one up and two down.

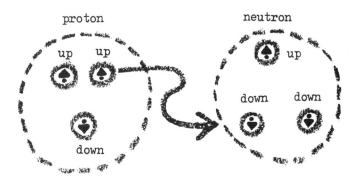

Transmuting a proton into a neutron thus involved changing an up quark into a down quark. The successful burning of the Sun and the making of the chemical elements depended on the ability of the weak force to accomplish what this diagram shows. But quarks were not alone in feeling the weak force, which shared with gravity the distinction of acting on matter of all kinds. It affected even the ghostly neutrinos, which have a special place in the story of the weak force.

Meet the neutrinos

Billions of neutrinos are passing in all directions through your body, as you read these words; by the time you pause to think about them, they have gone on their way, out past the orbit of the Moon. They are harmless; indeed they are undetectable, because they scarcely interact at all with the ordinary matter of the universe. But theorists came to realise that the Big Bang with which the universe evidently began must have scattered vast numbers of neutrinos, to wander completely at random through space, practically indifferent to the presence of the galaxies. And the burning and explosions of stars created more neutrinos: every time the weak force, the Cosmic Alchemist, changed one quark into another it produced a neutrino.

The existence of neutrinos was predicted in 1930 by Wolfgang Pauli. He needed them to explain why a radioactive atomic nucleus, shooting out an electron, lost more energy than was represented in the electron itself. Some uncharged particle was stealthily removing part of the energy – hence the idea of a 'little neutral one', which was what 'neutrino' meant. Without mass and without charge, a neutrino was plainly a curious particle indeed. It was an electron without electricity; it had energy and a certain spin, but that was all its heritage. Yet it was not a force-carrying particle which could casually appear and disappear again. Nature would count the numbers of neutrinos, and they ranked as durable particles.

A neutrino would feel the force of gravity, but because it travelled at the speed of light any deflection would be slight. Neutrinos colliding with the Earth would pass straight through and come out on the other side. Indeed, a neutrino could travel through many millions of miles of rock without feeling anything. Apart from gravity, the neutrino would react with ordinary matter only by the weak force, so its inertness testified to the extreme weakness of the weak force.

A quarter of a century elapsed after Pauli's prediction, before the first direct evidence of the neutrinos' existence was forthcoming. In 1956 Fred Reines and Clyde Cowan detected a very small number of the copious neutrinos produced by a nuclear reactor. They managed to 'persuade' them to alter the constituents of a few atomic nuclei. Other experiments suggested that there were two distinct types of neutrinos, one of which could join in reactions involving ordinary electrons, and the other associated with heavy electrons. In 1962 Leon Lederman and Melvin Schwartz from Columbia University used the accelerator at the Brookhaven National Laboratory to prove the existence of the two types of neutrinos.

They demonstrated that neutrinos released during the formation of heavy electrons could go on to produce other heavy electrons when reacting with nuclear matter, but never any ordinary electrons. After eight months' work and two million pulses from the Brookhaven accelerator, with each one showering large numbers of neutrinos through ten tons of aluminium, the experimenters had clocked-up about 30 cases of heavy electrons being produced. That was another example of how rarely neutrinos reacted with anything.

The ordinary electron, the heavy electron and the two neutrinos were all basic constituents of the universe – nature's small change. The physicists called the electron family the 'leptons', after Greek farthings. Just as there were anti-electrons, so there were anti-neutrinos, but one could call all the variants 'neutrinos' except where the distinctions became important.

A last general remark about neutrinos before we see how they revealed a new form of the weak force: they were an object lesson in how *not* to build an interesting universe. If all particles were as inert as the neutrinos and all forces as feeble as the weak force, the contents of the universe would wander as aimlessly and unconstructively as the neutrinos do, like microscopic Flying Dutchmen.

The picture of Nobody

'The neutrino was waiting for Gargamelle,' a senior physicist declared with gratitude when France formally handed over a new 'bubble chamber' as a gift to the high-energy physics laboratory at CERN near Geneva. Three French laboratories, at Saclay, Orsay and Paris, had spent six years designing and building it. André Lagarrigue of Orsay, who had led the French initiative, predicted results Rabelaisian in their scope.

In the sixteenth-century story by François Rabelais, Gargamelle was the mother of Gargantua, the gluttonous giant. She gave birth to the brat after imprudently eating sixteen hogsheads, two barrels and six jugs of tripe. The instrument installed at CERN in 1971 and honoured with her name (which the physicists had already jargonised to 'GGM') was excellently gross. It weighed a thousand tons all told and its belly contained ten tons of heavy liquid (freon) with an exceptional appetite for neutrinos.

European physics was waiting for Gargamelle, too. The CERN laboratory had long been a magnificent model of international research. The governments of Western Europe, in a collaboration older and broader than the Common Market, had created a better equipped laboratory for high-energy research on the particles of matter than any one country was willing to pay for alone. Young men from nations recently at war formed international teams to build big machines and carry out experiments with them. Yet supreme success eluded them.

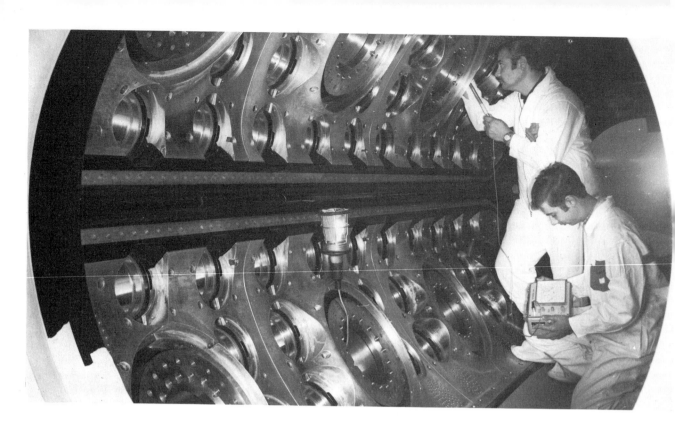

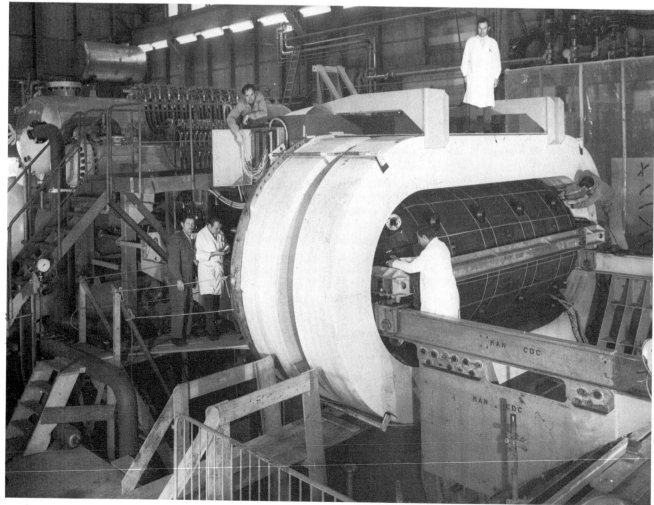

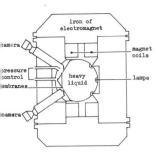

The centre-piece at CERN in the 1960s was a proton accelerator that served in many worthy experiments. But there was a very similar machine at Brookhaven, with which the Americans scooped several big discoveries. One could speak of knowledge belonging to all mankind, but it was discouraging for European physicists to have to confirm discoveries from across the Atlantic.

By 1970 the situation was changing. CERN was completing a new machine called the Intersecting Storage Rings which the Americans were not attempting to rival. And then there was Gargamelle. It was built in the belief that important new findings would come from close scrutiny of the elusive neutrinos. That was the reason for the heavy liquid, which took the place of the liquid hydrogen more usual in bubble chambers. The scheme was unfashionable, but Lagarrigue insisted on going ahead with Gargamelle despite open scepticism on both sides of the Atlantic. He died before his instrument's harvest of discoveries was fully apparent.

After its original invention at Berkeley nearly twenty years earlier, the bubble chamber became the physicists' favourite instrument for finding out how particles of matter behaved. Any electrically charged particles passing through it left plainly visible tracks, which you could photograph and analyse. They were rather like the vapour trails, or contrails, drawn across the sky by high-flying aircraft. But they were made by the boiling of a liquid, rather than by its condensation.

Essentially a bubble chamber is a tank of liquid maintained close to its boiling point under pressure; when the pressure suddenly eases, bubbles tend to form in it. A pump working with a repeatable action (differently engineered for different bubble chambers) reduces the pressure in synchrony with pulses of particles coming from an accelerator, and then restores it, ready for the next pulse. Wherever an electrically charged particle barges through the liquid, tearing electrons out of atoms and creating a disturbed region, the bubbles appear most readily.

With a well-judged decompression and a flash of light illuminating the liquid, experimenters can photograph the trails of small bubbles left in the wake of the particles. Set out before you are sudden appearances, disappearances and ricochets, as particles react with one another. Using more than one camera gives a three-dimensional view of the events. To provide further information a powerful magnet, the bulkiest part of the instrument, straddles the bubble chamber. It has the effect of bending the tracks sideways, steering positively charged particles one way and negative particles the other. The reluctance or readiness with which each particle swerves gives information about its mass and energy.

The obvious snag about using bubble chambers to study neutrinos was

Gargamelle: the heavy-liquid bubble chamber at CERN, Geneva, has made several major discoveries about cosmic forces and the qualities of matter. In the interior of the chamber (upper photograph) the large holes are for cameras; the small holes are for pressure control. In operation, the chamber is filled with a heavy liquid. Decompression of the liquid coincides with the passage of neutrinos through the chamber and any electrically charged particles set in motion by the neutrinos create lines of bubbles, which reveal their movements. In the exterior view (below) the light-coloured protuberances are the coils of a big electromagnet, which makes the particles swerve, thus giving information about their charges and momentum.

41

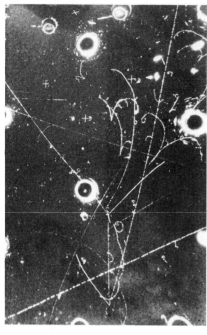

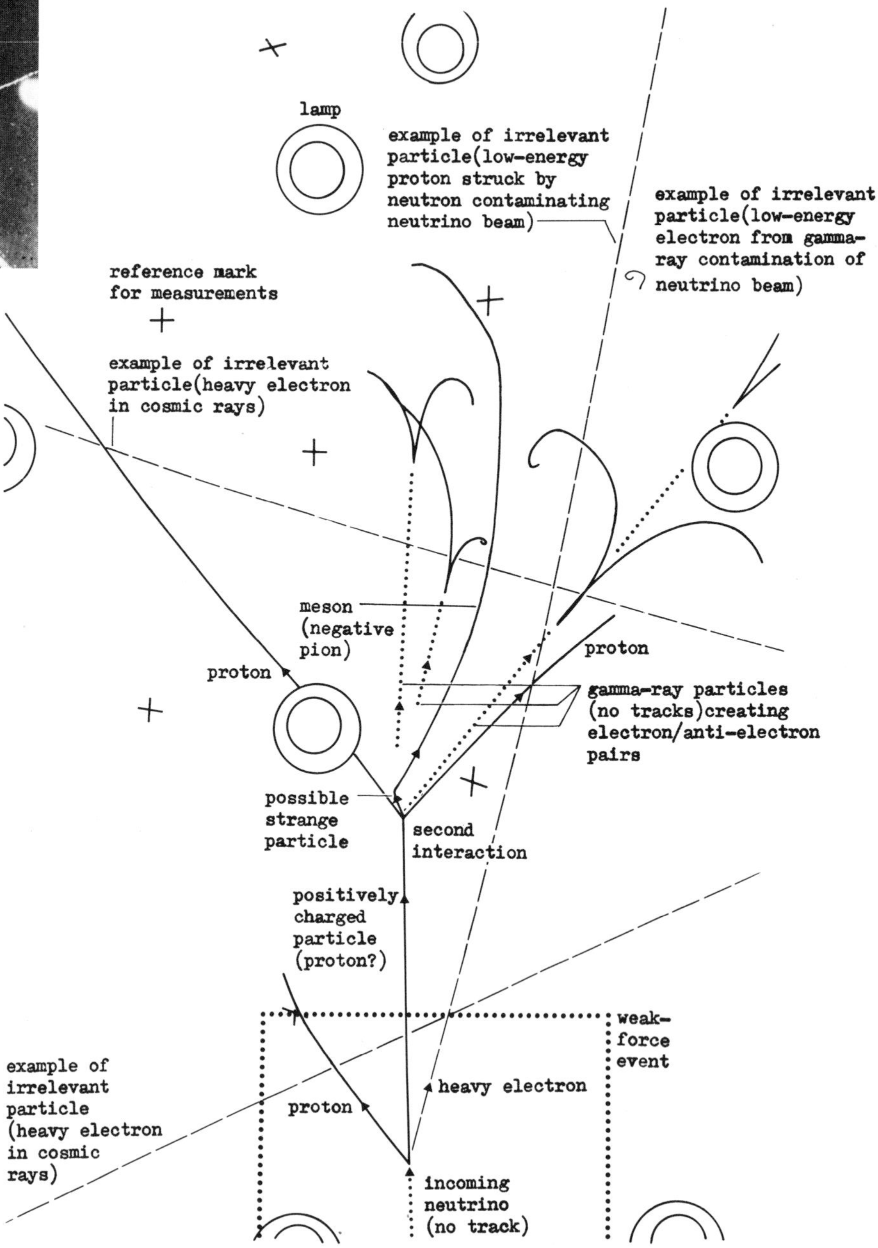

lamp

example of irrelevant
particle(low-energy
proton struck by
neutron contaminating
neutrino beam)

example of irrelevant
particle(low-energy
electron from gamma-
ray contamination of
neutrino beam)

reference mark
for measurements

example of irrelevant
particle(heavy electron
in cosmic rays)

meson
(negative
pion)

proton

proton

gamma-ray particles
(no tracks)creating
electron/anti-electron
pairs

possible
strange
particle

second
interaction

positively
charged
particle
(proton?)

weak-
force
event

A typical event recorded by the Gargamelle bubble chamber. Many invisible neutrinos are entering from the bottom of the picture. In the area boxed near the bottom, one neutrino reacts with an atomic nucleus in the heavy liquid, by the weak force. The neutrino changes into a heavy electron and the reaction sets other particles in motion. Cosmic rays and other particles produce stray tracks and make the picture look more complicated. The large blobs in the photograph are the shaded lamps used to illuminate the tracks.

example of
irrelevant
particle
(heavy electron
in cosmic
rays)

proton

heavy electron

incoming
neutrino
(no track)

that, having no electric charge, neutrinos themselves would leave no tracks. Instead, the experimenters had to look for the rare occasions on which a neutrino set other particles in motion, which did leave tracks. The greater the mass of liquid in the chamber, the more chance there would be of a neutrino passing closely enough to another particle to interact with it by the weak force.

So Gargamelle was provided with its ten tons of freon, and the new bubble chamber began pumping away among the pulses of neutrinos made by the proton accelerator at CERN. High-energy protons guided into a massive target created a spray of sub-atomic particles of various kinds. Most of them were shortlived and broke up, eventually making neutrinos and various by-products. For creating pure pulses, a massive shield was needed, which would absorb all particles except the neutrinos. The principal shielding in front of Gargamelle was a wall of iron 22 metres thick.

Each pulse of the accelerator propelled about a billion neutrinos through the bubble chamber. Synchronised with the pulses, Gargamelle would ease its pressure; its lamps flashed and its cameras took pictures of the chamber's interior to record any tell-tale trails of bubbles. As expected, most of the pictures showed nothing of interest. Even in Gargamelle, neutrinos rarely did anything. Because of the very short range of the weak force, a neutrino had to score almost a direct hit on one of the very small basic particles – quarks or electrons – of which the liquid was composed. The experiment was like trying to kill flies by firing a machine-gun in a pitch-dark forest.

The experimenters working with Gargamelle and analysing its pictures were looking at the general behaviour of neutrinos. The neutrinos were ideal for investigating the weak force. The rest of the particles available could react with other matter by the strong nuclear force and/or the electric force – both of which were so strong as to mask almost completely any effects of the weak force. You might have to wait patiently for a neutrino to react but, when it did, you knew you were seeing only the weak force.

According to the conventional view of the weak force, a neutrino reacting with anything had to change into another particle. For example, a neutrino made by the break-up of a heavy electron coming from the accelerator target would change back into a heavy electron whenever it scored a hit in the bubble chamber. Thereafter it would leave a conspicuous track. But a theory current in the early 1970s said that there could be a quite novel version of the weak force – a form never seen before. It would allow a neutrino to react with other matter and yet remain a neutrino.

The new version of the weak
force revealed by a Gargamelle
photograph. An invisible
neutrino enters and leaves the
bubble chamber unchanged.
Yet during its passage it reacts
with an atomic nucleus and
sets particles in motion,
principally a proton and a
negative pion, which is a
common type of meson, or
carrier of the strong nuclear
force. The pairs of tracks
labelled e/e are electron/anti-
electron pairs created as by-
products from very energetic
particles of light (gamma-rays).
Positively charged particles
curve to the left in the
magnetic field and negatively
charged particles to the right.

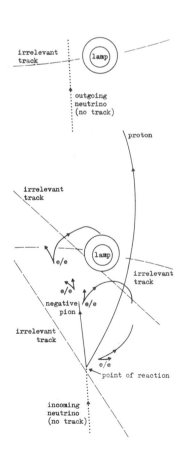

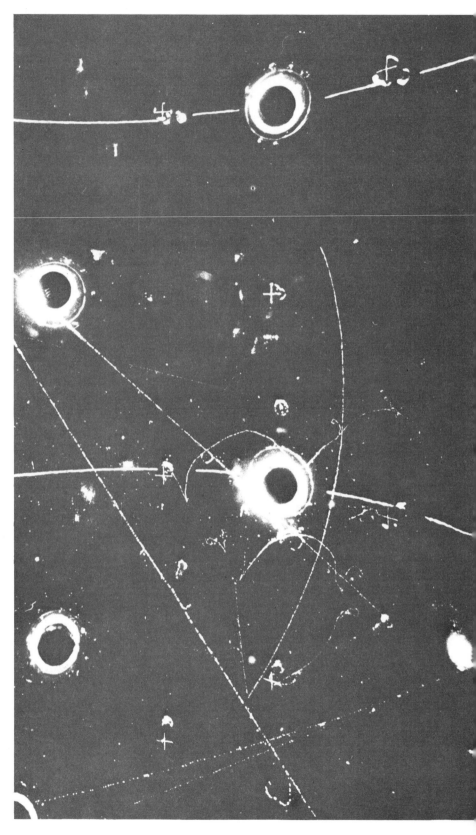

There were grave doubts about that possibility. If such a form of the weak force existed, then other kinds of particles ought to be able to break up with its help, in distinctive ways. Physicists had looked for such behaviour and found that it did not occur. Nevertheless, the Gargamelle experimenters decided to see whether by any chance their neutrinos would show this novel force in action.

It was, in its magnitude and delicacy, an heroic experiment to attempt. The crucial point about the novel force was that an invisible neutrino should enter the chamber, react with other matter and go on its way, remaining an invisible neutrino. 'This is the tune of our catch, played by the picture of Nobody,' Trinculo complained in *The Tempest*. The European physicists had to deduce the passage of a neutrino, and the exertion of the force, from the behaviour of other particles which the neutrino disturbed – particles which did leave tracks. And when they reckoned how many pulses in Gargamelle they would have to record in order to find sufficient evidence for or against the novel force, the answer was about a million – each one yielding a set of photographs that someone had to study carefully.

A special international collaboration was arranged for the purpose of examining the Gargamelle pictures. Some remained for analysis at the CERN laboratory itself; the others were distributed to laboratories scattered around Europe: in Aachen (Technische Hochschule), Brussels (Interuniversity Institute for High Energies), Paris (École Polytechnique), Milan (University and I.N.F.N.), London (University College), Orsay (Linear Accelerator Laboratory), and Oxford (University). I make no apology for the catalogue. I am telling of a major discovery, accomplished by many people working in a team; I am only sorry that I cannot name the 55 physicists concerned. Prominent among them were Donald Cundy, Paul Musset and André Rousset of CERN, Jean Sacton of Brussels, Ettore Fiorini of Milan, Fred Bullock of London and Donald Perkins of Oxford.

A new force in nature

Go into a laboratory where the bubble-chamber photographs are being analysed, and you will find technicians working in pools of light in a darkened room. Each of them sits at a big table on which the pictures can be projected one by one. Various mechanical and electronic aids help him or (more often) her to cope with the enormous numbers of pictures. Typically, two technicians scan every picture independently, to avoid individual oversights. Experts go through the pictures marking the events of special interest on tracing paper.

Then the tracks in question have to be measured with high precision. The technicians move a device about the table which automatically signals its whereabouts to a computer, whenever the technician is satisfied with the positioning and presses a button. The computer may complain if a succession of points along a track does not make sense to it.

Such is the patient work to which the laboratories of the Gargamelle Collaboration committed themselves. But the effort was rewarded. In about one picture in a thousand the new force showed up, with a neutrino reacting without changing its identity.

It was like discovering magnetism, or seeing gravity throwing an apple *sideways* off a tree – a great event in its own right, quite apart from its theoretical implications. By September 1973 the collaborators had accumulated enough cases to announce the new cosmic force, at a conference in Aix-en-Provence. The CERN press communiqué justly described it as 'an astonishing discovery'.

The novel cosmic force caused much less stir than, say, the resignation of Spiro Agnew (an American vice-president, you may recall) which occurred at about the same time. It did not help matters that the physicists would refer to it as 'the weak interaction via the neutral current', which sounded insipid as well as complicated. If, instead of mountains on the Moon, Galileo had reported 'superficial immobile irregularities', who would have listened to him?

And my name for it, Starbreaker? During the 1960s the theory of how a big star could explode had run into difficulties. It seemed clear that at a critical moment, with its nuclear fuel almost exhausted, the dying star would suddenly fall together under gravity. Then, the star released its last big supply of energy, with heat in abundance to fire a huge outburst of radiation. There was also pressure in abundance to squeeze the material at the centre of the star into an extremely dense state. But that was not all.

An exploding star really exploded. It broke, scattering most of its contents far and wide into space, thereby making the chemical elements available to later generations of stars – and to us. Some enormous force first checked the mighty fall under gravity and then drove the outer layers of the star spacewards at high speeds.

A possible factor that might account for it was the production of very large numbers of neutrinos, both by the final nuclear reactions and by the squeezing of the core. The neutrinos would act on the other matter of the star, by the weak force. In its old guise as the Cosmic Alchemist, that force was altogether too weak to accomplish the remarkable reversal involved in breaking the star. Then the new form of the weak force became 'respectable'.

Other laboratories found
similar evidence for the new
version of the weak force. This
bubble-chamber photograph is
from the Argonne National
Laboratory near Chicago.
Here the invisible neutrino has
dislodged a neutron (also
invisible) and a positive pion.
The latter particle identifies
itself by breaking up in two
stages into an anti-electron,
which leaves the conspicuous
corkscrew track.

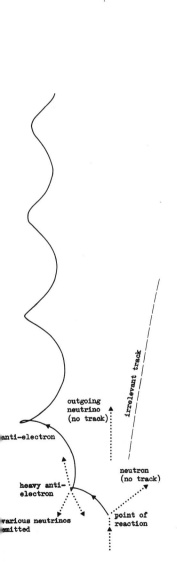

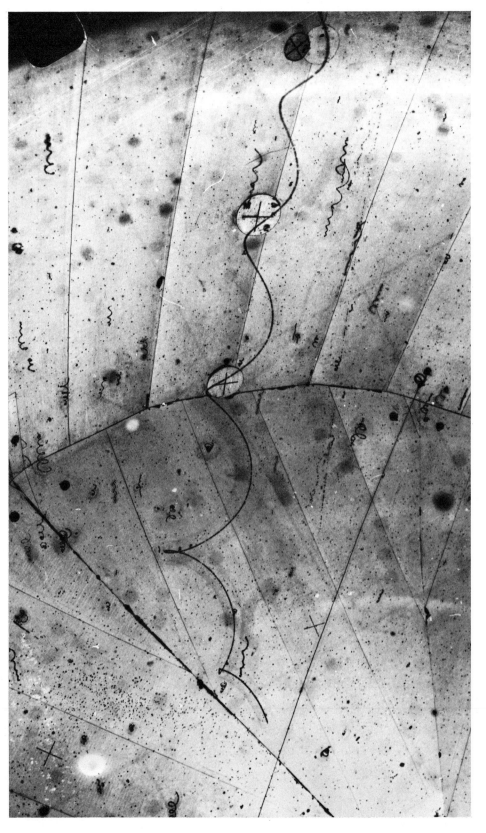

Not a three-armed instrument, but one seen in two large mirrors. It was installed in the neutrino beam at CERN by experimenters from Aachen and Padua to detect the new version of the weak force, manifesting itself in ways different from those normally seen in bubble chambers. The instrument is a bank of 'spark chambers' in which passing particles leave trails of sparks. They can be photographed with the aid of the mirrors.

Like the old form, it was inherently weak, yet in the unusual circumstances of an exploding star the neutrinos could use it to show exceptional 'muscle'. Daniel Freedman (State University of New York at Stony Brook) and James Wilson (Lawrence Livermore Laboratory, California) offered that explanation. Comparatively heavy atomic nuclei such as silicon and iron formed in abundance in the last stages of nuclear burning in the star. As Freedman pointed out, the swarming neutrinos could now interact with the heavy atomic nuclei, exerting a pressure on a nucleus as a whole which would be much stronger than the effect of a neutrino on an individual quark. When Wilson did the calculations, he concluded that the force was sufficient to break the star. The calculations were admittedly tricky and other astrophysicists preferred to look to magnetism or centrifugal forces for the disruption of the star. But the protagonists for the neutrinos and the novel force were confident that they had identified at least a major factor in stellar explosions. That was the connection between those hard-sought lines of bubbles in Gargamelle and the extraordinary events among stars that made our existence possible.

There was a second phase of discovery, in the CERN-based research into the behaviour of the Starbreaker. In virtually all of the events identified in the early runs with Gargamelle, the neutrinos were reacting with nuclear matter. An even more convincing demonstration of the new force would be to see a lone electron suddenly set in motion by the passage of an invisible neutrino. In 1973 only one such event had been found in 735,000 Gargamelle pictures. By 1976 Helmut Faissner and his collaborators from Aachen and Padua were claiming the detection of more than a score of them in a special investigation using particle-counters. Fred Reines and his colleagues in the US reported similar behaviour by neutrinos coming from a nuclear reactor.

Gargamelle was to play an important part in another exacting search – for charm, a new quality of matter comparable with electric charge. It was badly needed if physicists were really to say they understood the weak forces, and especially why the Starbreaker had not shown up earlier. But the story of charm is postponed to a later chapter. For the moment, I want to tell how some theorists predicted the novel version of the weak force.

Electricity and the weak force

A crazy idea which developed very gradually in the 1950s and 1960s was that the weak force might be the electric force in disguise. On the face of it, it was absurd. Could there really be any connection between the familiar deeds of the electric force, like lightning flashes and the swinging of

49

compass needles, and the nuclear alchemy in stars and radioactive materials? In any case, the electric force was unlimited in range, and involved no change in the electric charges on which it acted. The weak force was far weaker than the electric force and extremely limited in its range; also it did change the charges. To suggest they were akin was like comparing a jet plane with a gas cooker. Yet there were theorists willing to say so.

It began with the idea that you could visualise the weak force being carried by a force-carrying particle, in much the same way as a particle of light carried the electric force. We had this diagram earlier.

Electric force – a reminder

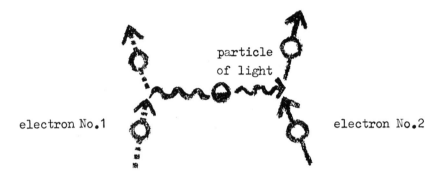

Similar diagrams might represent the weak force. Because the weak force in its role as Cosmic Alchemist was very versatile, there would be various diagrams. Here is one example.

Weak force – one case

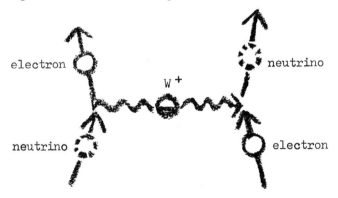

It shows a neutrino (neutral electron) emitting a force-carrying particle labelled W^+ (W for weak) and changing into an ordinary electron. The force-carrying particle then reacts with an electron and changes it into a neutrino. (The process would be exactly reversible, with an oppositely charged force-carrying particle, W^-, going the other way.)

The essential action of the weak force in stars, for allowing hydrogen to burn and elements to build, looked a little different.

The next diagram shows an up quark throwing out the force-carrying particle (W^+) and changing into a down quark. But now the force-

**Weak force
– another case**

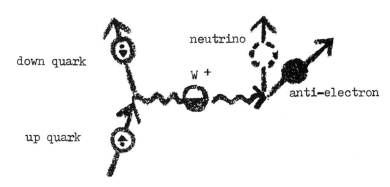

down quark

neutrino

W⁺

up quark

anti-electron

carrying particle is itself quickly breaking up, into a neutrino and an anti-electron. The diagram contains interesting clues about the nature of the W⁺ particle. For a start, the up quark has changed into a down quark. How could that trick be done? My first chapter told how nature could create short-lived particles – particle and anti-particle – out of nothing, as it were. It used 'borrowed' energy, permitted by the uncertainty about how much energy was present at any instant. Imagine that there was created in the neighbourhood of the up quark such a pair of particles: a down quark and anti-down quark. Next, imagine the down quark replacing the up quark (this is the alchemy we are after). The up quark could then join up with the remaining anti-down quark to make the W⁺ particle, thus reconstructing the left-hand side of the previous diagram.

**A 'borrowed'
pair helps in
the alchemy**

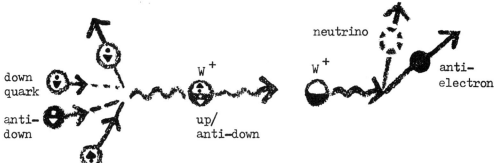

down
quark

anti-
down

W⁺

up/
anti-down

neutrino

W⁺

anti-
electron

But on the right-hand side we saw the W⁺ particle breaking up into a neutrino and an anti-electron, and nothing else. So we should be equally entitled to say that the W⁺ particle *consisted* of a neutrino and an anti-electron. There was the powerful 'magic' of the weak force as Cosmic Alchemist. The force-carrying particle could change its spots as it travelled along – and change back again.

**The W⁺ particle
varies its constitution**

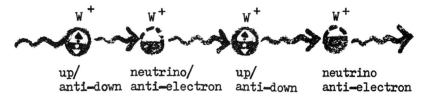

W⁺   W⁺   W⁺   W⁺

up/   neutrino/   up/   neutrino
anti–down   anti–electron anti–down   anti–electron

There were other possibilities for its constitution, which need not concern us now. And by taking the opposite constituents, you could have a very similar W⁻ particle, going through the same spot-changing routine.

The W⁻ does likewise

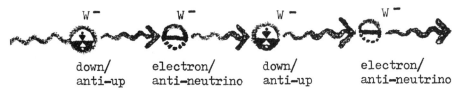

| down/ anti-up | electron/ anti-neutrino | down/ anti-up | electron/ anti-neutrino |

This variability enabled the W particles to react with all kinds of matter, and it meant also that any one constitution was like any other. It showed a hidden kinship between apparently quite unrelated particles, such that the *difference* between a down quark and an up quark was somehow the same as the difference between an electron and a neutrino.

The W^+ and W^- force-carriers carried electric charges and could account for the well-known activities of the weak force, known before 1973. But now suppose there were another force-carrier, with no electric charge, denoted W^0. Because there would then be no obligation for the particle which emitted it to change its electric charge, it would allow a neutrino, for instance, to react with a quark, say, and still remain a neutrino. In other words, that was the new form of the weak force which I have called the Starbreaker. And the variable constitution of the W^0 particle might then be thought of as follows.

The W⁰ particle also changes its spots

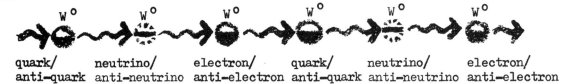

| quark/ anti-quark | neutrino/ anti-neutrino | electron/ anti-electron | quark/ anti-quark | neutrino/ anti-neutrino | electron/ anti-electron |

And in that diagram is the nub of the theory, because it includes an electron and anti-electron combination. In the previous chapter we said that, with caution, one could think of a particle of light, the carrier of the *electric* force, being composed of an electron and anti-electron. So if you imagined a particle of light behaving very oddly, and changing its spots as indicated in the last diagram, then it could become a carrier of this novel form of the weak force. Moreover, if you imagined the W^0 particle reacting to a particle in a suitable way, you could make the W^+ (or W^-) particle, like this.

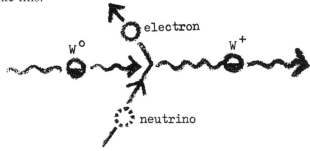

In other words, the weak force could be understood as the electric force in disguise, if particles of light could go berserk from time to time, like a freak giant wave in a storm at sea. Nor would the unusual behaviour have to occur very often, because the weak force was very weak compared with the electric force. If only one force-carrying particle of light in a thousand behaved oddly, it would account for the weak force and all the cosmic alchemy of which it was capable.

The three hurdles

The conundrum of the electric and weak forces took time to solve. The description just given of the working of the weak force involved very odd behaviour by light-like particles. It allowed them to change their constitution, both in reacting with other particles including uncharged particles, in ways that ordinary light did not, and by private changes occurring in mid-flight. Three formidable hurdles had to be surmounted before this idea could be taken at all seriously.

First, someone had to explain why the weak force only operated at very short, sub-nuclear ranges, while the electric force had unlimited range. As we shall see, that meant the W particles carrying the weak force were heavy – but why were they heavy?

The second hurdle had to do with the calculations of the strength of the force. There seemed to be too many ways in which light-like particles carrying the weak force could operate. By that reckoning, the force should be infinitely strong, rather than weak!

The third and most interesting hurdle was that this theory necessarily predicted a version of the weak force that no one had yet seen – tantamount to a novel cosmic force. In order to reduce the two forces, electric and weak, to one you had first to increase them to three, by adding the weak force Mark II, represented by W^0.

That the W particles had to be very heavy had been apparent for many years. When Enrico Fermi offered the first theory of the weak force in 1934, he assumed their masses to be infinite – in effect, not force-carriers at all. The extremely short range over which the weak force acted implied a short lifetime for the force-carriers and therefore a heavy mass. The reasoning here linked up with the uncertainty about energy, which allowed the 'borrowing' of a little energy for a comparatively long time, or a lot of energy for a very short time. Later estimates put the W particles at perhaps 50 to 70 times heavier than a proton, or as heavy as an iron atom – so their lifetime and range were correspondingly short.

As a particle of light had no true mass, anyone trying to claim that the

W particles were like light faced an enormous discrepancy in mass. As early as 1954 Chen Ning Yang and Robert Mills, working at the Brookhaven National Laboratory near New York, developed some key mathematical ideas that might in principle unify the forces acting between particles. But their theory dealt with particles without mass. The Yang-Mills theory was very tantalising, but it could not be directly applied to any real forces in nature. Although able theorists kept returning to the problem, notably Sheldon Glashow in 1961, and Abdus Salam and John Ward in 1964, ten years passed before there was any definite progress towards unifying the electric and weak forces.

It was by no means obvious that the idea was really worth pursuing. Always, after a breakthrough in ideas, one might look back and see something like a logical development of theories preceding it, with some individuals sticking doggedly to the task. But at the time, hundreds of other bright ideas were circulating, most of them doomed to come to nothing. Military tacticians speak of the 'fog of war'; it is transparent compared with the fog of fundamental physics in the making, where the direction in which one should be facing is not even approximately clear.

The discrepancy in masses, between the particle of light on the one hand and the heavy W particles on the other, was a severe case of 'broken symmetry'. A more familiar case is the way a person always writes with one hand rather than the other, which 'breaks' the left/right symmetry of the human body. The social custom of shaking hands with the right hand is also a broken symmetry.

The weak force exhibited a similar and very remarkable kind of broken symmetry. The discovery was a great sensation of the 1950s, but it was called 'parity violation' which made it sound harder to visualise than it was. The simple and startling point was that nature knew the difference between right and left.

Most of the particles of the micro-universe were known to possess spin – that is to say, they acted as if they were like little planets spinning about their axes. Until the mid 1950s, nature seemed to be entirely equitable about which way a particle might be spinning. If you thought of a particle spinning in the manner of a rifle bullet around its line of motion, it was just as likely to spin to the right (clockwise) as to the left (anti-clockwise).

**Equality (parity)
of left and right**

That was nicely symmetrical, and fitted with general notions about the perfection of the micro-universe. Then some Chinese-born physicists working in America pursued the idea that maybe, when the weak force went to work, it broke this particular symmetry. One of them was Yang, co-author of the Yang-Mills theory already mentioned. Appropriate experiments were done, and it turned out that when a radioactive material threw out electrons (by the weak force) they were much more likely to be spinning to the left than to the right.

This could be interpreted as meaning that the associated neutrino *always* spun to the left, which was confirmed experimentally.

In anti-matter, the bias was reversed. Richard Feynman nicely illustrated the meaning of this discovery by speaking of a rendezvous with an alien spaceman. How do you check that he is not made of anti-matter, which could be disastrous for you both? You tell him in advance to do the experiments in radioactivity which distinguish our left from our right. You tell him about our social traditions, too. Then you meet him and, as Feynman said: 'If he puts out his left hand, watch out!'

Biologists, for their part, wondered whether a broken symmetry in life on Earth, the 'left-handedness' of the molecules found in all living organisms, might have something to do with the bias in the weak force. Theoretical physicists later came to suspect that a clearer understanding of why the bias occurs would come out of the theory that unified the weak force and the electric force.

The make-weights

The working of the electric force, too, showed cases of broken symmetry. One was the ordinary compass needle, which had a 'north' and 'south' pole, so you could not say it was the same either end on. A more spectacular but also subtler broken symmetry occurred in superconduction, when metals cooled to very low temperatures lost all resistance to the flow of electric current. Both of those effects were produced by the action of electrons, whose behaviour in normal circumstances was quite symmetrical.

In the early 1960s two young British physicists, Jeffrey Goldstone and Peter Higgs, made important theoretical discoveries. They realised that whenever nature spontaneously broke (some preferred to say 'hid') the basic symmetry of a system, it created peculiar new vibrations or even new particles. They developed the concept of rather mysterious particles of a new kind, associated with the broken symmetry between the particles of light and the W particles, carriers of the weak force. The crucial point was that W particles would, as it were, swallow these new particles and become heavy, while an ordinary particle of light, by spurning them, stayed 'slim'.

First hurdle: how could the W particle put on weight?

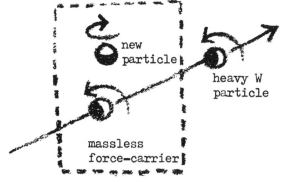

The left-handedness shown by the weak force meant that this coupling of particles could occur only for left-handed force carriers: only they could make the W particles and carry out the operations of the weak force.

In 1967 Steven Weinberg of the Massachusetts Institute of Technology seized on this idea to advance a simple and concise theory unifying the electric and weak forces, which put the particle of light and the W particles into the same package. Independently Abdus Salam, working at Imperial College in London, and the International Centre for Theoretical Physics in Trieste, arrived at the same scheme. The first hurdle, the need to account for the discrepancy in the masses of the force-carrying particles of the electric and weak forces, seemed to be crossed. Yet this Weinberg-Salam theory aroused very little excitement.

One good reason for scepticism was that it was still not clear how the theory could avoid predicting a weak force of infinite strength, for that is how their mathematical calculations always seemed to end up. Both Weinberg and Salam gave their opinion that this, the second hurdle, ought to be negotiable, but were unable to say precisely how to do it. The unification of the forces languished for a few more years.

Steven Weinberg, originator in 1967 of the standard theory uniting the weak force and the electric force. He was then at the Massachusetts Institute of Technology but he later moved down the road to a joint appointment at Harvard University and the Smithsonian Astrophysical Observatory. His interests span gravitation and cosmology as well as particle physics (see Chapter 6) and he is much concerned with problems of arms control as well. Weinberg was born in 1933.

At Utrecht University, a systematic programme of theoretical research was under way at the beginning of the 1970s, trying to find a method of calculating the weak force in ways that gave sensible answers. Whenever you took account of the fact that two particles could pass more than one force-carrying particle between them (as they certainly had to be allowed to do) the force seemed to become impossibly strong.

Second hurdle: how to calculate the force correctly?

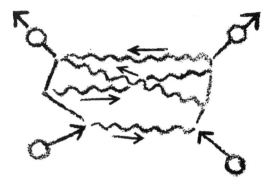

Diagrams like these were a theorist's nightmare. But the man leading the work at Utrecht, Martin Veltman, had made progress. He was a skilled manipulator of the diagrams. He could add diagrams together and re-present many of them with just one. He could subtract them, too, and in some cases one dreadful infinity would neatly cancel out another. But in other cases the force remained stubbornly infinite, as a computer duly confirmed to Veltman.

One of his research students, Gerard 't Hooft, suggested that Veltman was not assuming quite the right sort of particles. Veltman was sceptical at first, but he tested 't Hooft's proposals on the computer, and it told him point-blank that 't Hooft was right. Sensible answers came out of the theory for the first time. It was one of those rare moments when people struggling up a mountain of ideas suddenly found themselves at the summit, with a whole new scene laid out before them.

What 't Hooft did was to bring together ideas about how nature could operate – making, as it were, its own calculations of the force without the benefit of a computer, in a workable system. You had to have all the pieces for it to function.

As 't Hooft commented to me:

'With hindsight things look so obvious, you don't understand how they missed it all. . . . Veltman had the mathematical technique but the wrong idea about the particles involved. Weinberg and Salam had the right idea about heavy force-carrying particles, but not the technique to do the computations, so nobody took it seriously. Some Russians had relevant tricks, but they hadn't exploited them. I came fresh into it, and just combined the right things.'

The news from Utrecht sent people hurrying to their libraries to look

Abdus Salam, who arrived independently at the same theory as Weinberg. He directs the International Centre for Theoretical Physics at Trieste, as well as being professor of theoretical physics at Imperial College, London. He was born in 1926, in a rural district of British India, in what is now Pakistan. His manifest ability took him to Cambridge University to complete his education. He has been in the thick of the action in theoretical physics ever since, and prominent in UN work.

up those old papers by Weinberg and Salam. It also sparked off a flurry of activity in generating further theories about the cosmic forces. The second hurdle had been crossed. Yet, in 1971, for the ideas unifying the electric and weak forces, the third hurdle remained, in the form of the variation of the weak force with the neutral force-carrier W^0. The theory said that neutrinos could react with other particles, and disturb them, while remaining neutrinos themselves.

Third hurdle: a new kind of force in nature?

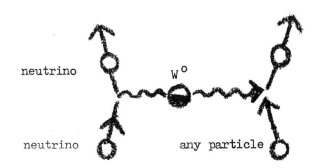

That was exactly what the Gargamelle Collaboration discovered in their photographs in 1973. Other laboratories including Fermilab and the Argonne National Laboratory in the USA confirmed it. If the going had been easier, the outcome might have been less impressive. As it was, three difficult hurdles had been cleared, in transforming a crazy idea into a computable theory that was attested, at least provisionally, by experiment. Nature had been obliging the whole way down the course presumably because, in spite of all the early appearances, the weak force was indeed the electric force in disguise. The most striking proof of kinship would be to alter the speed of radioactive decay, or the break-up of particles, using a very powerful magnet. Experiments were put in hand to test this possibility. It was completely alien to previous ideas, according to which radioactivity followed its own laws, quite independently of external influences of any kind. Further confirmation of the unified theory would have to wait until machines more powerful than any available in the 1970s could make observable specimens of the very heavy W^+, W^- and W^0 particles themselves, and so demonstrate their reality.

Gerard 't Hooft of Utrecht University who, as a 24-year-old research student, showed that theories of the kind that unite the weak force and the electric force could give sensible answers. That was in 1971. Since then he has developed the so-called 'gauge theories' further, for example in showing that they predict a heavy magnetic monopole (so far undiscovered) which Paul Dirac also predicted many years before by different reasoning.

The gauge theories

The theory that united the weak and electric forces was of the kind known as a 'gauge theory'. The surge of hope after 1971, that the forces and particles of the universe might indeed be comprehensible to the human mind, came about because young Gerard 't Hooft showed how to make gauge theories work well enough to give sensible answers. One reason for the excitement was that these theories had a special status, as the kind of thing that you would expect in a well-ordered universe.

To understand what the physicists meant by a gauge theory, it may be helpful to think of a piece of decorated wallpaper and the kind of symmetry which it represents. It is quite different from, say, the simple left/right symmetry of the human body. The same pattern is repeated over and over again. The task facing the designer of wallpaper is to make sure that complicated patterns will repeat reliably when the paper-hanger joins two lengths of wallpaper side by side.

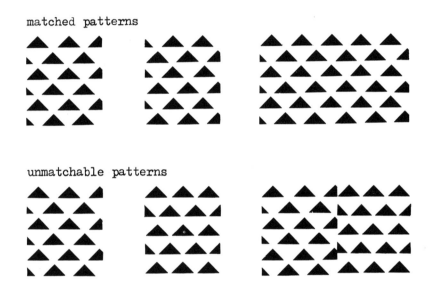

Properly designed paper has consistency across an expanse of wall. The 'designer' of a universe would have a similar problem of ensuring consistency. For example, if the laws of nature were different from one galaxy to the next, you might ask: where is the 'join' in space where the laws change, and would not very peculiar things happen there? The prime discoveries of astronomy could be summed up by saying the universe was like wallpaper.

Hydrogen atoms, for example, evidently behaved in just the same way in the most distant galaxies as they did on Earth; gravity and the nuclear forces, too, were plainly conforming to familiar laws, so that those distant galaxies looked very much the same as our own Milky Way. But the distant galaxies were also 'a long time ago'. Because of the time it took for their light to reach the Earth, astronomers saw them as they had been millions or billions of years ago. So there was consistency, or symmetry, in time as well as space.

Or think of an ordinary stone. To say that it remained the same from

59

one day to the next, and would not alter if you carried it from place to place in your pocket, was like saying it turned up unchanged at different points in space-time: the pattern of the stone repeated reliably, anywhere in the cosmic wallpaper. Matters of symmetry in space and time therefore had to do with the very nature of existence, and the basic principles of a viable universe.

Nature had no special laws for the safeguarding of stones. Instead there were laws that conserved particular qualities of a stone (or any other object) which together guaranteed its existence and persistence. For instance, a law of conservation of momentum prevented a stone from spontaneously jumping out of your pocket and flying away to the Moon, while a law of conservation of electric charge forbade the atoms of the stones to fall apart. You might imagine a crumbly universe in which such laws did not apply, but there would be no long-lived observers able to admire the chaos.

For physicists, to speak of symmetry in nature or of laws of conservation amounted to the same thing. To give another example: the symmetry between electrons and anti-electrons was in part an expression of the conservation of electric charge. You could create a charged particle out of available energy, but only if you created an oppositely charged particle at the same moment.

The crucial feature of the gauge theories was the way in which they confronted, head-on, the issue of the consistency of the laws of nature throughout space-time. They therefore promised to be an important part of any key to the universe. In particular, the gauge theories related one piece of the cosmic wallpaper to another, by considering how local events were connected with distant events.

The first fully-fledged gauge theory was the modern account of the electric force. One well-known characteristic of the electric force was that the force between any two charges was not affected by how many other charges there were around. This meant, for instance, that you could charge a whole laboratory up to a high voltage and your measurement of the force between two electrons would be quite unaffected. To put it another way, the theorists were at liberty to specify the circumstances for which they chose to calculate the local effects of one electron on another. They called this choosing the gauge.

But how could you have a consistent universe if any small piece of it could be considered independently, as far as the electric force was concerned? Two features of the universe took care of it: first, the conservation of electric charge; secondly, the ability of electric charges in different parts of the universe to communicate with one another.

The modern theory of the electric force got rid of mysterious ideas about action at a distance, in which one electric charge attracted or repelled another without any discernable connection between them. It said that 'virtual' particles of light passed between electric charges, carrying the force. And the passage of these particles of light guaranteed that any discrepancies between one part of the universe and another would be automatically ironed out. For example, if you were doing your experiments in a laboratory at a high voltage, and then tried to relate yourself to the rest of the universe, the forces acting on your laboratory would quickly tell you that everything else was at a different voltage.

If electric charges could appear and disappear, these comparisons would not be reliable and terrible inconsistencies would arise. Indeed, the whole argument could be inverted to say that electricity in a consistent universe ought to involve conservation of electric charges and the passage of force-carrying particles between them. Plainly it would be reassuring if other cosmic forces could be described in similar ways, especially if it brought out a deep-lying kinship between new forces.

The electric force was comparatively simple and elegant to describe in this way, because the force-carrying particles had no mass and travelled at the speed of light; also they had no electric charge themselves. Theorists ran into difficulties when they tried to apply the same gauge idea to the weak force and the strong nuclear force. They wanted to assert the greatest possible symmetry in the workings of these cosmic forces: to say that regardless of how you chose to describe the forces at any one place, the way in which the forces acted would automatically ensure consistency with events in any other place.

An obvious snag was that the force-carriers for the weak force were not nearly such perfect messengers as particles of light. They were massive and short-lived, and did not travel at the speed of light. Nevertheless, it turned out that consistency could be achieved provided the force-carrying particles felt the force themselves. For forces of that kind, the basic Feynman diagram became a little more complicated.

'Gauge theory': force-carrying particles feel the force themselves

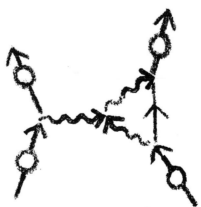

Two particles approaching each other could both send out force-carrying particles. But the force-carriers could meet in the middle and feel each other's effects – which modified their behaviour.

This forked diagram summed up better than any other single picture the way fundamental physics was going in the 1970s. It represented gauge theories in general and all the cosmic forces could be described by it, in a way that suited the goal of self-consistency in the universe. It also accorded perfectly with the idea that the force-carrying particles shared the qualities of the particles on which they acted: so much so that they were themselves vulnerable to the force. The electric force fitted into the general scheme, as just a special case in which the interactions between the force-carrying particles were negligible until they 'went berserk' and operated as the weak force.

The progress in understanding the weak force, described in this chapter, flowed from the gauge idea. The turning point came when the mathematics of gauge theories proved to be manageable. Before that, theorists had wondered whether they could make a gauge theory for the strong nuclear force. Afterwards they did, but indirectly, by going to the very heart of things and considering the forces between the quarks of which nuclear particles were composed; and there they found the mother of the strong nuclear force.

The Cosmic Forces

Gravity

(gravitation)
chief cosmic roles: binding planets, stars,
    galaxies
extremely weak in atoms; very strong in
    collapsed stars ('Starcrusher')
infinite in range
acts on everything (matter and energy)
carried by gravitons

Weak Force Mark I

(weak interaction via charged current)
chief cosmic role: altering basic particles
    ('Cosmic Alchemist')
strength about $1/_{100,000,000,000}$ of electric force
very short in range (10^{-15} centimetres)
acts on all basic particles (quarks and electron
    family)
carried by W^+, W^- particles

Electric Force

(electromagnetic interaction)
chief cosmic roles: binding atoms; creating
    magnetism
strong in atoms; weak over cosmic distances
    because matter is neutral
infinite in range
acts on all charged particles
carried by photons

Weak Force Mark II

(weak interaction via neutral current)
chief cosmic role: effective in exploding stars
    ('Starbreaker')
strength about $1/_{100,000,000,000}$ of electric force
very short in range (10^{-15} centimetres)
acts on all basic particles (quarks and electron
    family)
carried by W° particles

Strong Nuclear Force

(strong interaction)
chief cosmic roles: binding atomic nuclei;
    burning in stars ('Sunfire')
about 100 times strength of electric force
short in range (10^{-13} centimetres)
acts on quarks
carried by mesons

Colour Force

(chromodynamic interaction)
chief cosmic role: binding quarks in protons;
    strong nuclear force is a by-product
extremely strong but weak at short ranges
in practice, short in range (10^{-13} centimetres)
acts on quarks
carried by gluons

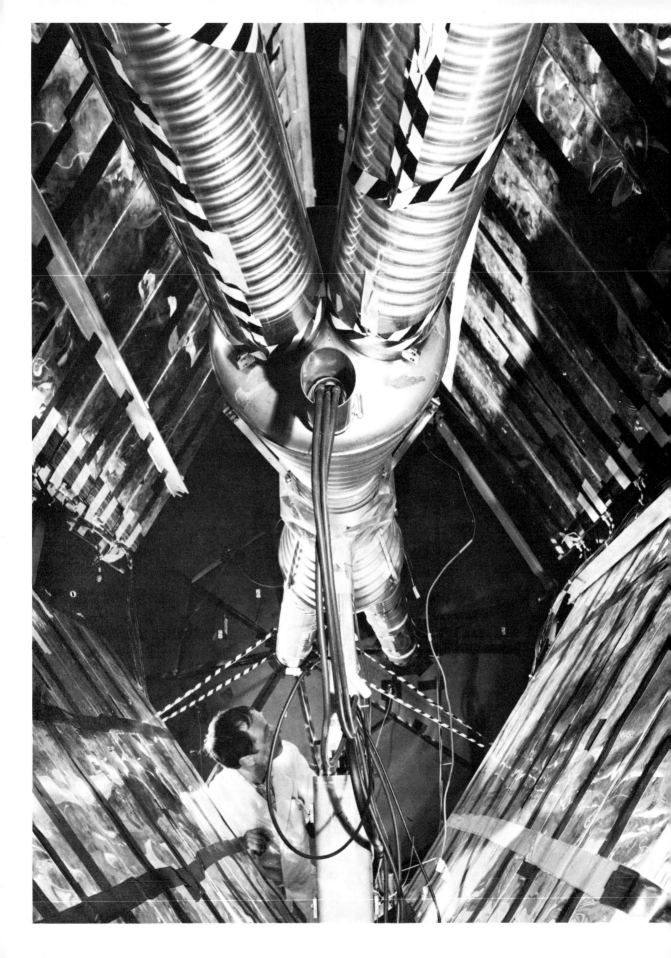

3 Sunfire

All of the cosmic forces turned out to be fire of a kind. Conventional chemical fire, for example in the blazing logs of the first human firemakers, released energy by the work of the electric force, regrouping atoms and drawing them closer together. The weak force was a cold sort of fire, except in its supporting role in the stars. Nevertheless the radioactive decays which it helped to engineer on Earth, in radium for example, released rays that could create a glow in material exposed to them.

Gravity as 'fire' might seem a conundrum, until you stopped to think of incandescent meteorites streaking through the air, converting their fall to Earth into frictional heat. If they reached the ground they could set forests ablaze. And at the birth of an ordinary star, the falling together of a cloud of gas created quite enough heat to make the star extremely hot and bright.

Nineteenth-century physicists supposed, in fact, that gravity was all the fire the Sun had, releasing heat by binding the stuff of the Sun more closely together. Hermann von Helmholtz saw it as a continuous process, in which the Sun gradually shrank, releasing new energy all the time. Calculations on those lines by William Thomson (Lord Kelvin) indicated that the Sun was 20–30 million years old. He thought that was satisfactory, but already there was evidence from the geologists of fossil remains of life considerably older than that.

By the mid-twentieth century it was plain that the Sun was more than a hundred times older than Thomson supposed. Nor was it shrinking at any significant rate: the heat released at its core created pressures that resisted gravity. But by then a more durable source of fire was known – the strong nuclear force. While gravity would draw matter in general together, or chemical fire bind atoms more tightly, the strong nuclear force, the true Sunfire, gripped the nuclear matter in the heart of atoms.

Energy of binding

The closer binding of matter by any kind of force meant a release of energy, equal to the energy you would have to expend to prise it apart again. When Albert Einstein discovered that mass was equivalent to

ntriving the collision of
tons, the prime heavy
terial of the universe, in
RN's Intersecting Storage
gs. At this cross-over point
wo vacuum vessels, protons
velling in opposite
ctions meet head on and
ct with great violence.
ticle-counters surrounding
intersection record the
ducts of each impact and
p to show what protons are
de of.

energy, it did not only tell physicists that matter was a kind of frozen energy. It also implied that the release of energy by any means would reduce the mass of material involved. Tightly bound material would always be actually lighter than loosely bound material, even if the difference was hard to measure when the forces were gentle.

Regardless of what trick the Sun used to burn by, Einstein's equation ($E = mc^2$) required that it be losing more than four million tons of mass every second, in pumping out light and heat at the observed rate. It became clear that the strong nuclear force enabled the Sun to go on losing mass at that rate for far longer than any other force could allow.

The reduction of mass achieved by the release of energy during chemical binding was proportionally very small. For instance, the molecules of water produced by combining hydrogen and oxygen would be only very slightly lighter than the sum of the hydrogen and oxygen before their combination. If the Sun were made of hydrogen and oxygen and burned by conventional chemical means, it would have to consume millions of billions of tons of its fuel every second to achieve the necessary loss of mass-energy.

In burning hydrogen in the nuclear fashion, with the aid of the strong nuclear force, the Sun's fuel consumption was less than a billion tons a second, for the same output. That was insignificant compared with the total mass of the Sun, which could therefore go on burning for billions of years. Nor was it hard to see where the loss of mass occurred.

From the hydrogen fuel of the Sun four protons, nuclei of hydrogen atoms, could go into building one nucleus of helium, the next heavier element. When physicists weighed these nuclei precisely, in laboratories on Earth, there was a plain discrepancy. Helium was lighter than four protons, so there was the loss of mass, supplying the energy. The binding of the pieces by the strong nuclear force resulted in so great a release of energy that the mass was reduced by more than half of one per cent. Later calculations showed that more than a quarter of the energy was lost, in the form of inert neutrinos.

Matter into energy

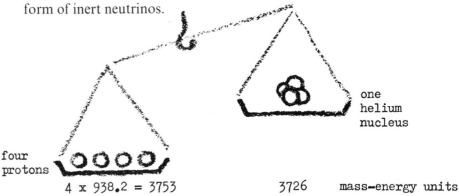

one helium nucleus

four protons

$4 \times 938.2 = 3753$

3726   mass-energy units

The nuclear energy that human beings soon came to release on Earth was also the energy of binding, by the strong nuclear force. This Promethean operation, whether manifested in bombs or in power-generating reactors, depended on the special properties of certain materials available at the Earth's surface. Uranium was the heaviest chemical element that occurred in any noticeable quantities. Exploding stars manufactured heavier elements; so did scientists on Earth. Apart from some very heavy oddities discovered in 1976, they were usually much less durable than uranium, which was itself on the margin of elemental survival.

The reason why the nuclei of heavy atoms were unstable was plain: the strong nuclear force between their constituents was only about a hundred times stronger than the electric force. Uranium, with 92 electrically charged protons in its nucleus, was nudging at the limit, beyond which the force of electric repulsion between those protons would break up the nucleus. In radioactive decay, discovered at the end of the nineteenth century by Henri Becquerel, uranium could gradually rid itself of its excess protons by throwing out small pieces of its nucleus. With less opposition from the electric force, the strong nuclear force could bind the constituents more tightly, with a corresponding release of energy. It transpired that uranium changed down through a succession of chemical guises until it finished as the metal lead, one of the heaviest of the stable elements, with 82 protons in its nucleus. But radioactivity was, at best, a source of very limited power for special purposes.

Uranium showed a more dramatic and energetic method of breaking up, which opened the way to the large-scale release of nuclear energy on Earth. The uranium nucleus could split into two big pieces, with a great release of energy. This nuclear fission was unexpected and the announcement of its discovery by Otto Hahn and Fritz Strassmann in Berlin came in January 1939, at a moment which unhappily guaranteed that the first use of the new source of energy would be military.

The uranium nucleus, especially in its rarer, lighter form (U-235), could be prodded into fission by a neutron – the uncharged relative of the proton and itself an important constituent of nuclear matter. But the break-up produced, besides the two large pieces, two or three neutrons. They could go and cause other nuclei to split, producing more neutrons, that would power a bomb or a controlled reactor. It also turned out that the artificial element plutonium, which could be made from uranium in a nuclear reactor, could also sustain a fission chain reaction. The Hiroshima bomb used U-235, the Nagasaki bomb plutonium. I leave aside the history and politics of nuclear energy, to attend to the cosmic forces involved.

67

The reason why much more energy came from the gross splitting of a uranium or plutonium nucleus than from its radioactive decay was that elements of about half the weight of uranium possessed far fewer protons in the nucleus. The disruptive electric force was weaker and the strong nuclear force had a much tighter grip. So, when a uranium nucleus split, the ingredients of the two main fragments drew closer together; the binding released energy and abolished a corresponding amount of mass. The resulting fragments were radioactive because they now had too many neutrons. The weak force duly corrected the proportion.

In one way or another all of the cosmic forces were involved in the release of the energy of nuclear fission on Earth. The heavy atoms had been manufactured in exploding stars, and gravity gathered them in building the planet. The fierce tussle between the electric force and the strong nuclear force made the heavy nuclei unstable and released the intense energy. Finally the weak force created the obnoxious radioactivity of the 'fission products'.

Because nuclei in the middle of the table of elements were more tightly bound by the strong force than the heaviest or the lightest, you could release binding energy, starting from either end of the scale. Nuclear fusion, as developed for the H-bomb, was more directly akin to the fire of the Sun. The H-bomb burned heavy forms of hydrogen to make helium.

Starting from the lightweight elements, intense heat was required to overcome the electrical repulsion between the nuclei and let the strong nuclear force establish its grip. In the core of the newborn Sun, gravity supplied the necessary heat, which thereafter came from the energy released by the nuclear burning itself. To achieve comparable temperatures, the H-bomb required a detonator in the form of a fission bomb.

Less drastic ways of achieving nuclear fusion on Earth for peaceful purposes were sought in many countries. A variety of techniques to bring about collisions of sufficient energy between the nuclei of heavy-hydrogen fuel seemed promising, but a practical fusion reactor was slow to appear. Meanwhile, the H-bombs in the armouries of five nations constituted the dismal pinnacle of human mastery over the cosmic forces.

Patterns among the particles

While all that was going on, fundamental research continued in its more detached mission of trying to understand how the strong nuclear force actually worked. I have already mentioned (p. 28) the brilliant theory of Hideki Yukawa in Japan, who said that the constituents of the atomic

Hideki Yukawa, founder of the modern theory of the strong nuclear force. He was born in 1907 and in 1934 was working at Kyoto University when he developed the idea of mesons as the force-carriers. Later, particles were discovered that fitted his prescription exactly, although only after a long period of confusion in which the heavy electron was mistaken for Yukawa's meson. Yukawa won the Nobel Prize in 1949.

Murray Gell-Mann of the California Institute of Technology, who has done more than any other individual to bring order out of sub-nuclear chaos. He was born in 1929 and he made his first important mark on physics in 1953, when he named 'strangeness': the quality of matter which enabled some particles to live longer than expected. By 1961 he was grouping strange and non-strange particles in mathematical families. Gell-Mann suggested in 1963 that the imaginary objects shuffled in the mathematical manipulations were real objects, which he called 'quarks'. Other theorists were developing similar ideas at each stage, but Gell-Mann's contributions were exceptional and won him the Nobel Prize in 1969.

nucleus exchanged force-carrying particles, in much the same way as particles of light carried the electric force. Yukawa's 'mesons', as they were called, proved to be real. In fact, after 1947 there was quite a plague of different kinds of mesons turning up, more than Yukawa ordered or the perplexed physicists would wish for. And they were only part of the sub-nuclear chaos that was setting in.

In the early 1930s the contents of the universe seemed simple. From just three kinds of particles, electrons, protons and neutrons, you could make every material object known at that time. Thirty years later human beings were confronted with a bewildering jumble of dozens of heavy, apparently elementary particles, mostly very short-lived. They came to light either in the cosmic rays or in experiments with the accelerators.

The particles had various mass-energies and differing qualities such as electric charge, spin, lifetimes and so forth. Moreover they were given confusing, mostly Greek, names, so that one of the most eminent of physicists, Enrico Fermi, was driven to remark before his death in 1954: 'If I could remember the names of all these particles I would have been a botanist.'

The proliferation could be understood to some extent, in that many of the particles seemed to be energetic relatives of the proton. Because they possessed greater inherent energy their masses were greater. Each was in some sense less tightly bound together than a proton and it could quickly change into a proton with a release of binding energy and an associated loss of mass. But that implied the proton was not a truly basic particle: it was made of something else which could be bound more or less tightly together.

A small group of theorists brought order out of the chaos. The principal figure among them was Murray Gell-Mann of Caltech, then in his early thirties. He declared that all the heavy particles of nature were made out of three kinds of quarks. He had the word from a phrase of James Joyce: 'Three quarks for Muster Mark.' It was a mocking cry of gulls which Gell-Mann took as referring to quarts of beer, so he pronounced quark to rhyme with 'stork'. Many other physicists rhymed it with 'Mark'. In German, as sceptics were not slow to notice, 'quark' meant cream cheese, or nonsense. But the name was far less important than the idea.

Great credit was due to other people, notably Soicho Sakata at the University of Nagoya and Yuval Ne'eman from Israel. George Zweig, also of Caltech, arrived at the theory of quarks independently of Gell-Mann and called them 'aces'. But while these others came up with crucial ideas at the right moment, Gell-Mann was in the thick of the theoretical action through all the creative years around 1960, and as a result came to dominate the scene.

The greatest puzzle in the menagerie of particles was what, in the end, brought enlightenment. Some of the heavy relatives of the proton, bursting with excess mass-energy, were uncommonly slow to change down into protons. They had lifetimes of around one ten-billionth of a second – very short by human standards yet more than a million times longer than expected. They were called 'strange' particles and they were said to represent a new quality of matter called strangeness.

The discovery of strangeness in the early 1950s was more unexpected and revolutionary than most of the discoveries occurring afterwards – many of which flowed from the new thinking about nuclear matter which strangeness demanded. Curiously little credit went to the cosmic-ray physicists – Clifford Butler, 'Goku' Menon, 'Beppo' Occhialini, Cormac O'Ceallaigh, George Rochester – who found the oddly-behaving particles that spoke of the new quality of matter. Gell-Mann named it 'strangeness' in 1953.

What did strangeness mean? You could compare it with electric charge, a very familiar quality of matter because of all the experiments with charges and charged particles. Electric charge meant in essence a quality of particles which caused them to behave in predictable ways, in response to the electric force. But it was also accountable, in that each electron, for instance, had the same electric charge, and you could count the number of charges. In fact, even if electric charge could not be detected just by rubbing a stick on your sleeve, physicists could have discovered it in experiments with particles. They would have found that *something* imposed a rule about what kinds of products could appear from a reaction between particles. The sum of the electric charges coming out always had to equal the total charge going in.

Strangeness was a quality affecting the lifetimes of particles. Unlike an electric charge it did not survive indefinitely – indeed the eventual break-up of strange particles into more ordinary particles involved the abolition of strangeness by the weak force. But strangeness persisted in processes involving the strong nuclear force and was strictly accountable. For instance, some particles had a double dose of strangeness, and in their behaviour you could always follow those two doses until they had both disappeared.

Bells began to ring in the theorists' minds when they realised that the strange particles bore a standard relationship to other particles. You could line up the normal particles according to weight and other qualities, and put the strange particles out to one side. Particles with a double dose of strangeness went further out. Then you finished up with families that made striking, highly symmetrical patterns. For example, here is the proton and its nearest relatives, with strange ones among them.

The proton and its family

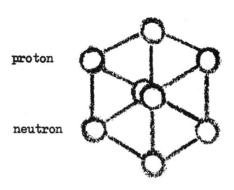

non-strange   strange   doubly strange

proton

neutron

Another family of particles consisted of 'souped-up' protons with considerably greater mass energy.

'Souped-up' relatives of the proton

non- strange doubly
strange   strange

This second pattern gave a broad hint to the theorists. They made the crazy prediction of a very peculiar particle with three doses of strangeness, which would complete the triangle by filling in the empty corner. Many of their colleagues refused to take the pretty pattern seriously. But in 1964 the particle in question, the omega-minus, turned up in a bubble-chamber picture at the Brookhaven National Laboratory. It possessed the predicted triple dose of strangeness. The pattern-makers won the day.

Quarks for shuffling

Then the sceptics were readier to listen to what Gell-Mann and others were saying about why the patterns existed. It would be wrong to imagine

71

The discovery of the omega-minus particle, appearing briefly near the bottom of the bubble-chamber photograph. This is one of the most celebrated pictures in the history of physics, because it confirmed that the patterns of particles being offered by Murray Gell-Mann and others had real meaning. It was predicted as a heavy particle that would survive long enough to leave a track in a bubble chamber, and would have three doses of 'strangeness'. In February 1964, Nicholas Samios and his colleagues at the Brookhaven National Laboratory found this picture among their bubble-chamber results.

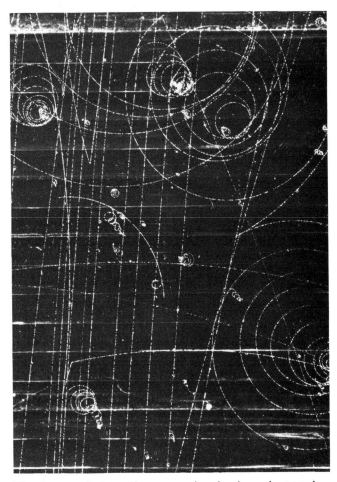

The photograph shows the omega-minus leaving a short track, breaking up in three stages (shedding one dose of strangeness on each occasion) and finishing up as a proton.

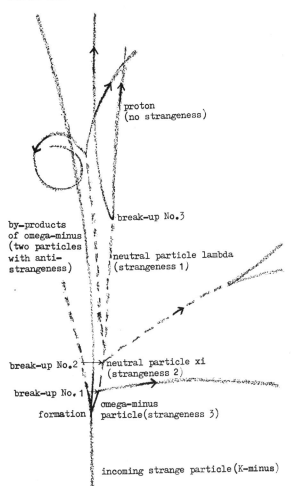

proton
(no strangeness)

break-up No.3

by-products
of omega-minus
(two particles
with anti-
strangeness)

neutral particle lambda
(strangeness 1)

break-up No.2

neutral particle xi
(strangeness 2)

break-up No.1

formation

omega-minus
particle(strangeness 3)

incoming strange particle (K-minus)

that the theorists sat around like children cutting out triangular or hexagonal patterns. There was a strong prop of mathematics and physics behind the patterns, in particular, 'group theory'. A young Frenchman, Evariste Galois, had introduced group theory into mathematics in the nineteenth century by hurriedly writing down all his ideas the night before he was killed in a duel. By the twentieth century, it was a fitting vehicle for deep ideas about the universe.

A group, in group theory, was any collection of objects, numbers or mathematical equations, related by any kind of repetitive process that changed one element in the group into another. For example, a very small group consisted of the three equal-sided triangles that could be made by rotating one triangle repeatedly through a third of a turn. All the possible arrangements produced by shuffling a pack of 52 playing cards made up a very large group. There were much subtler and more abstract things you could do with groups than that, but, among the particles of nature, the patterns of interest were generated by a combination of operations like rotation and shuffling.

The operations could also be expressed as the creation and annihilation of certain entities from which the particles might be made – in a word, the quarks. Creation and annihilation then meant substituting one quark for another. If you notice some peculiarities in the patterns, such as the presence of two particles in the middle of the proton's family, they are due to nature sticking to the rules of group theory rather than childish games.

By declaring there were three types of quarks, you could make all the protons and all its relatives by simply putting them into combinations of three. Gell-Mann called the three types of quarks u, d and s, for 'up', 'down' and 'sideways'. The implied directions had no physical meaning and, as 'sideways' alliterated with 'strange', people have since preferred to speak of the 'strange' quark. The quarks carried electric charges. The up quark was said to have a charge equal to two-thirds of the charge on the proton, while the down and strange quarks each had a negative charge of one-third.

Two up quarks and one down quark made the proton, the nucleus of hydrogen and the principal raw material of the universe. Three strange quarks made the famous omega-minus particle. The relatives of the proton all fell nicely out of the prescription, with the appropriate electric charges, for example:

proton	$\frac{2}{3} + \frac{2}{3} - \frac{1}{3} =$	1
neutron	$\frac{2}{3} - \frac{1}{3} - \frac{1}{3} =$	0
omega-minus	$-\frac{1}{3} - \frac{1}{3} - \frac{1}{3} =$	-1

Making particles by shuffling quarks

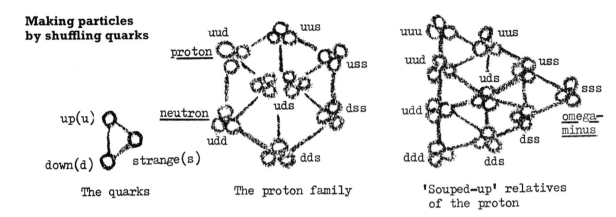

The quarks

The proton family

'Souped-up' relatives of the proton

'Upness', 'downness' and 'strangeness' all represented real and distinguishable qualities of matter, comparable in cosmic status with electric charge. Whether in experiments on Earth or in fierce reactions in the heart of the Sun, nature kept careful accounts of them. Each of the family names qualities implied a new law of nature. And each of the various particles composed from the quarks all behaved a little differently from one another, because of their different rations of these various qualities of matter.

There would be anti-quarks, too, forming anti-particles of all of these. For example, the anti-proton consisted of two anti-up and one anti-down (how tiresome those terms are!). Perhaps the most exciting feature of the whole scheme was that Yukawa's mesons, the particles that carried the strong nuclear force, were also made of quarks.

To be precise, each of the force-carrying particles was made of one quark in combination with an anti-quark. It therefore existed where the short-lived force-carriers belonged, in the no-man's land between the universe and the anti-universe. Here is one of the families of mesons which carry the strong nuclear force, and set the Sun ablaze.

Carriers of the strong force – the mesons

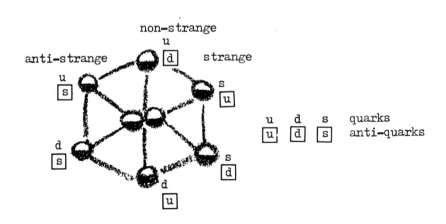

How the same combination of quarks can make different particles.

Each quark spins about an axis and the directions of spin of two quarks can be opposite or similar.

less energetic:
lighter particle

more energetic:
heavier particle

By convention, the spin of a quark is said to be '$\frac{1}{2}$', so two quarks spinning in opposite directions have a combined spin of 0; two quarks spinning in the same direction have a combined spin of 1.

Force-carrying particles

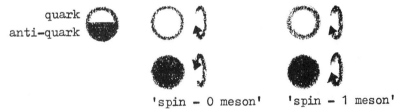

quark
anti-quark

'spin – 0 meson'   'spin – 1 meson'

Proton and relatives

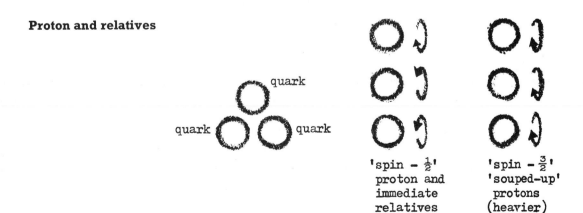

quark

quark   quark

'spin – $\frac{1}{2}$'
proton and
immediate
relatives

'spin – $\frac{3}{2}$'
'souped-up'
protons
(heavier)

Families of particles made
from three types of quarks and
their anti-quarks.

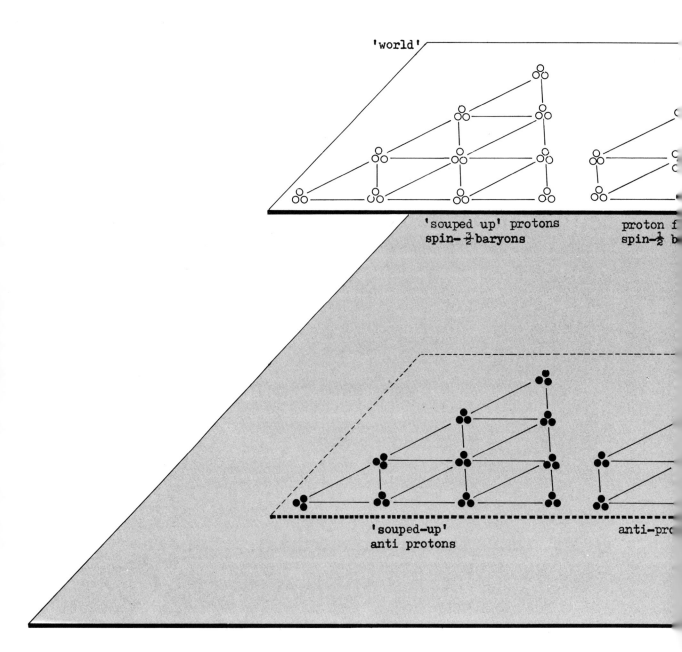

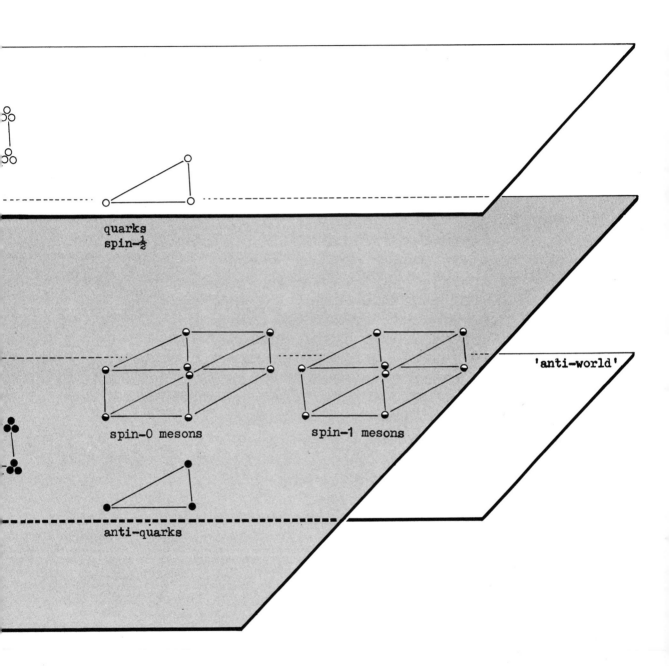

quarks
spin-$\frac{1}{2}$

spin-0 mesons

spin-1 mesons

'anti-world'

anti-quarks

Yukawa's scheme for binding the pieces of the atomic nucleus together could be reinterpreted in the light of the quark theory. One could see the force-carrying particles reacting with other particles whose qualities they shared. The strong nuclear force between a proton and neutron, for instance, became a force between individual quarks in the two particles, conveyed by a quarky meson.

The strong nuclear force as a game for quarks

proton   neutron

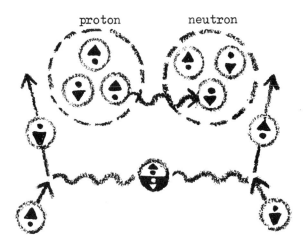

The essential simplicity and aesthetic appeal of the quark theory spoke for themselves. As we shall see later, it carried the seed of another train of thought that would put the Sunfire itself in a completely new perspective. But, after the success with the omega-minus, the immediate job for physics was to see whether the proton was really made of three quarks, as the theory said it must be.

Pieces in the proton

In years of effort that followed the development of the quark theory, no one managed to liberate a quark. All kinds of bizarre experiments were done, trying to discover free quarks lying around on Earth. But the quarks were eventually found in their lairs, using the great machines that could discover what protons were made of by peering right inside them. To penetrate the strong force-barriers guarding the protons' secrets required the use of probes of correspondingly enormous energy.

One such machine could accelerate electrons until each of them possessed 20,000 mass-energy units (20 GeV). The Stanford linear accelerator stretched for more than three kilometres across the Californian countryside – three kilometres of precision engineering creating a 100-million-dollar ski-run for electrons. Repeatedly down the line the electrons felt intense radio pulses which added to their energy. Because Stanford's

Using electrons to explore protons. The upper photograph, opposite, shows part of the long and straight electron accelerator at the Stanford Linear Accelerator Center (SLAC). It directed very energetic electrons into a target of liquid hydrogen. Massive detectors, including the one shown in the lower photograph, separated and classified the electrons scattered from the protons in the target. The details of the scattering implied that the electrons were encountering small, quark-like objects inside the protons.

78

machine was so straight and long, its pulses of electrons were the most energetic in the world, and the first to break into the inner sanctum of the proton.

The experimenters at Stanford fired electrons into a target of liquid hydrogen and, with suitably grandiose detectors, collected the electrons coming out on the far side. By measuring the angles and energy at which the electrons emerged, they could make deductions about the nature of the tussle between the electrons and the protons in the target. Electrons would not feel the strong nuclear force, but they could react by the electric force with any charged particles inside the protons – and quarks were said to be electrically charged.

In a classic series of experiments with the Stanford machine, it became clear that the proton was made of pieces. While some electrons simply swerved under the influence of the proton as a whole, others became involved in violent reactions which cost them energy. That happened too often, if the proton was thought of as a single entity, but the results made excellent sense assuming that the proton contained a little swarm of much smaller but heavy particles. An electron passing by chance close to one of the swarm could then feel a very strong effect.

At the CERN laboratory in Geneva, the Gargamelle bubble chamber, which figured so prominently in the story of the Starbreaker, was also giving evidence about quarks. When the neutrinos used in the Gargamelle experiments encountered quarks, in the liquid filling the bubble chamber, they could occasionally react with them by the weak force. Results forthcoming by 1974 again corresponded with the presence of a swarm of particles inside the proton. This experiment gave information on the characteristics of the proton's components that complemented the Stanford results.

Meanwhile CERN had brought its Intersecting Storage Rings into operation. This machine could subject protons to greater stresses than any other, because it collided two beams of them. Protons of up to 30,000 mass-energy units (30 GeV) circulated in opposite directions in two rings, almost a kilometre around. Magnets tamed the energetic protons and caged them in the vacuum chambers of the rings for hours on end.

At various points around the circle the rings intersected, allowing the protons to collide. As the collisions were almost head-on, they avoided the loss of energy by recoil. When a proton from a more conventional machine ran into a stationary target, any particle which it struck would tend to be knocked on, like one billiard ball hit by another; that moderated the violence of the impact. But in CERN's machine the colliding protons could, in principle at least, bring each to rest and thus make all their energy of motion available for reactions between the particles. The intersection-points in the storage rings were besieged with particle-counters that recorded the products of the collisions.

The first big discovery with the Intersecting Storage Rings was made and confirmed in different ways by several international research teams. When the protons collided, heavy energetic particles of various kinds frequently came out *sideways* – not at a small angle, such as you might expect from two 'fuzzy' protons glancing off each other, but even at right angles to the beams. Physicists were reminded of Ernest Rutherford's discovery of the atomic nucleus in 1911, when he found energetic particles bouncing off a metal foil. Only the close approach of very small, heavy objects could produce the big deflections of particles observed either in Rutherford's experiments or in the CERN Intersecting Storage Rings. Again it was strong evidence for pieces in the protons.

At Fermilab in the United States which possessed the biggest machine of them all, capable at its limit of accelerating protons to 500,000 mass-energy units (500 GeV), many of the elaborate experiments done with it bore in one way or another upon the quark theory. I shall just mention one of them which was in progress in 1976, Experiment 398. It represented a continuation by other means of the pioneering work with the Stanford electron machine. Physicists from Chicago, Harvard, Illinois and Oxford were using heavy electrons generated when the accelerated protons hit an intermediate target. With them they probed ever more intimately into the composition of the protons in a tank of liquid hydrogen. They were looking hard to make sure that the picture of the pieces of the proton, made out with Stanford's electrons, was no illusion but held good at very high energies.

As experimenters, they were paid to be sceptical, just as theorists were paid to have daring ideas. Although they knew they were probably seeing quarks, many experimenters insisted on calling the pieces of the protons, rather non-committally, 'partons'. They declared that it remained to be confirmed that the partons were quarks. Nevertheless, by the mid-1970s, nearly all the experiments seemed to be fitting the same picture of the proton.

Quark-like objects and glue

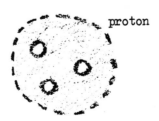
proton

Three heavy, electrically charged particles, presumably the quarks, sat fairly loosely in a sea of 'glue'. The glue accounted for an appreciable part of the mass-energy of the proton, and its behaviour implied that it consisted of a number of force-carrying particles, which acquired the name of gluons. In 1976 the physicists of Fermilab's Experiment 398 were able to report that this theory gave 'an attractive and complete explanation' of what they saw.

The glue as particles – 'gluons'

The colour force

Anyone tempted to sit back and just admire the pretty patterns of particles created by the quarks would be no theoretical physicist. Any advance in understanding a piece of the universe usually threw up new questions. The quark theory was no exception.

Why were only certain kinds of combinations of quarks permitted: three quarks as in a proton, or quark and anti-quark as in a force-carrier? Why not a pair of ordinary quarks, or four quarks in combination?

What about the glue needed to hold the quarks together inside the proton and the other permitted particles? What was the nature of the gluing force and the gluon particles that carried it?

What about lone quarks? If everything in the world was made of quarks, would they not be easy to find and isolate?

All of these questions were answered eventually (though not un-controversially) in a sweeping new theory. It prescribed the existence of a new cosmic force: a superstrong force, far surpassing the old strong nuclear force and in some ways quite unlike any force known before. It also involved a new quality of matter, again roughly comparable with electric charge. 'Colour' was the name given to the new quality of matter.

The remainder of this chapter which deals with the colour force (and also with a philosophical aspect of the present theories) is more speculative than most of what I am relating about particles and forces. At the time of writing the interplay of theory and experiment has not progressed very far and there is only limited or indirect evidence about the nature of the force between quarks. Yet, for the theorists who developed the idea of the coloured quarks, the pattern of particles and forces coming out of it made it a deeply satisfying and elegant theory that 'ought' to be correct.

20s:
very small
ject in the
art of a
drogen atom

50s:
small object
th mesons
its
cinity

60s:
fairly
rge object
ompared with
e electron)
nser at the
ntre than
the
ges

70s:
large
ject
ntaining
ch
aller
jects

eoretical
cture:
ree
arks
changing
uons

Of course, no one was suggesting that quarks were actually coloured, in any way resembling the colours to be seen in the large-scale world. But the term was apt, to the extent that a little knowledge of the behaviour of real colours provided helpful comparisons in the micro-world. All observable particles were said to be 'white', but the whiteness could be produced by mixing colours in various ways.

The notion of coloured quarks first appeared soon after the quark theory became established. Oscar Greenberg of the University of Maryland first speculated about colour in 1964. But the theory of superstrong colour force did not fully crystallise until the early 1970s, when progress with the gauge theories opened the way to a more thorough understanding of coloured quarks and coloured glue.

A reminder about real colours, from which the analogies were drawn, may be useful. If you mix coloured lights by adding them, as in a colour television set or on a spinning disc, you increase the variety of particles of light entering the eye. The rules and effects are a little different from the more familiar mixing of paints, which is a process that subtracts more and more kinds of particles of light out of the white light that falls on the paint.

In mixing lights, there are two standard ways of producing white. One is to add together three primary colours, conventionally red, green and blue. The other is to mix two colours, a primary colour and its complementary colour or 'anti-colour'.

colour	anti-colour
red	turquoise
green	mauve
blue	yellow

In these little lists, if you add three coloured lights downwards (red + green + blue or turquoise + mauve + yellow) you see white. Alternatively, if you mix any colour with its anti-colour, reading horizontally, you again see white. Any other mixture will produce a coloured, non-white effect.

The combinations of colours fitted neatly with the permitted combinations of quarks, assuming that nature insisted that the composite particles should look white. For example, you could say that a proton consisted of three quarks, one to be coloured red, one coloured green and one coloured blue. An anti-proton consisted of three anti-quarks with anti-colours: turquoise, mauve and yellow. In either case, the composite particle was white.

83

Colour force/colour light analogy

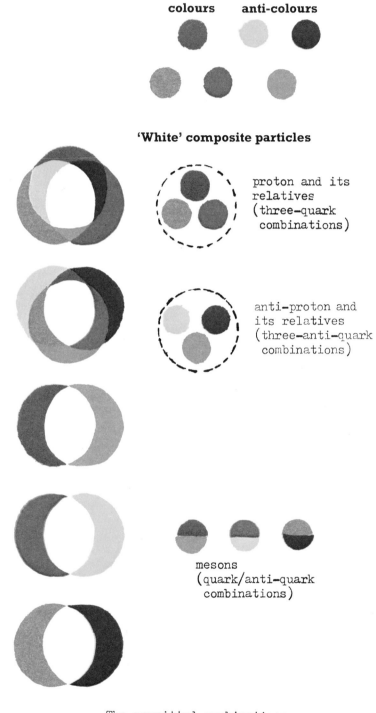

colours   anti-colours

Mixing coloured lights

'White' composite particles

proton and its
relatives
(three-quark
combinations)

Mixing anti-coloured lights

anti-proton and
its relatives
(three-anti-quark
combinations)

Mixing colours and anti-colours

mesons
(quark/anti-quark
combinations)

The permitted combinations
of quarks match the laws of
colour-mixing precisely. The
quarks are not really coloured.

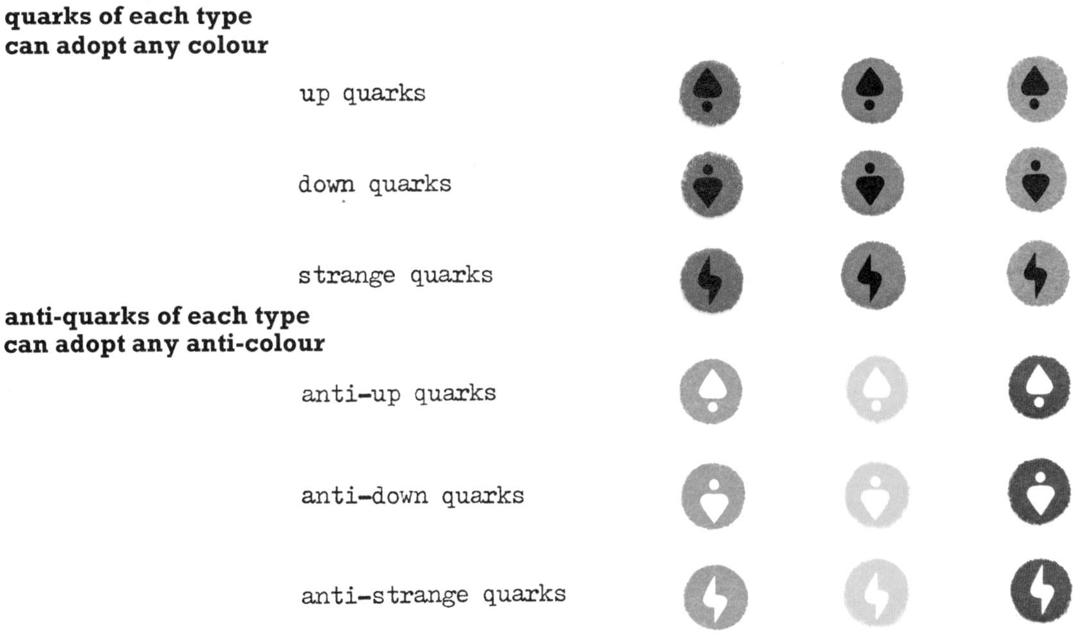

**quarks of each type
can adopt any colour**

up quarks

down quarks

strange quarks

**anti-quarks of each type
can adopt any anti-colour**

anti-up quarks

anti-down quarks

anti-strange quarks

Yukawa's mesons carrying the strong nuclear force consisted of a quark and an anti-quark. They could be said to look white to the outside world because a blue quark, for example, was matched with a yellow anti-quark. The assignment of a colour to a quark had no connection with what type of quark it was. A strange quark could be red, green or blue; so could an up or down quark. Indeed the theory of the colour force required that each quark in a proton was forever changing its colour.

And the force-carrying particles operating between the quarks, the gluons of the glue? They could each be thought of as drawing on the colours of a quark and an anti-quark, providing a mixture of colour and anti-colour. But because gluons did not show themselves in public, outside the proton, they could themselves be non-white. In other words, the colour did not have to be matched by its anti-colour and you could make a variety of coloured gluons:

red-mauve	green-turquoise	blue-turquoise
red-yellow	green-yellow	blue-mauve

In addition there were red-turquoise, green-mauve and blue-yellow (colours and their exact anti-colours) which might seem like three white gluons, but which for subtle reasons of group theory mixed together to make two 'off-white' combinations. That made eight kinds of gluons altogether, comprising a family very like those already encountered in the proton family and carriers of the strong force.

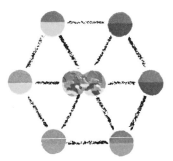

The force-carrying gluons would shuttle about between the three quarks of a proton, changing their colours. In practice, several gluons would be active at the same time, producing a scene like this, very like the picture of the proton emerging from experiments.

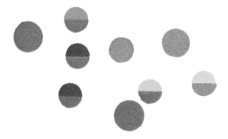

The operation of six 'charges', three colours and three anti-colours, compared with just two charges (+ and −) for the electric force, helped to ensure that the colour force would be very strong indeed. Nevertheless, it was a force of essentially the same kind as the other cosmic forces. We can represent it by a Feynman diagram similar to those we have used before.

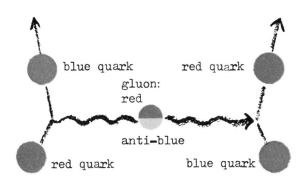

The gluons, being coloured, would themselves feel the colour force. Two gluons meeting could react, with an anti-colour in one abolishing colour in the other and so altering the precise version of the colour force that it carried.

Gluons feel the colour force

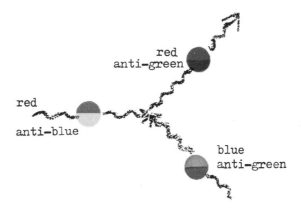

red
anti-green

red

anti-blue

blue
anti-green

The theory in which the force-carrying particles felt the force themselves was a 'gauge theory'. Thus the colour idea developed in the same way as the leading theories of the other cosmic forces. And it surged ahead after 1971, when Gerard 't Hooft had shown how to cure the calculations in gauge theories and arrive at sensible answers.

Freedom and slavery among the quarks

The fact that it was a gauge theory was the principal vindication of the whole hypothesis about coloured quarks and gluons. Without that feature, it might have been an absurd fantasy, a mere confection of coloured chalk on the blackboards in the theorists' offices, erasable all too easily. There was only indirect evidence of the existence of colour, or the charge-like qualities of matter to which that metaphorical name was assigned. But it became an uncommonly persuasive idea.

Once you declared it to be a gauge theory – that the force must feel the force – your options were instantly narrowed. Out of many possible conjectures about quarks and glue, and many conceivable patterns that families of particles might make, only a few were compatible with the decision to make a gauge theory. And among that remaining minority of possible schemes, the simplest of them demanded the symmetrical families of the protons and the mesons actually observed, and added the eight gluons of the colour force in a similar pattern.

Moreover, the picture that the gauge theory painted of the colour force made it out to be highly unusual, as cosmic forces went – but unusual in

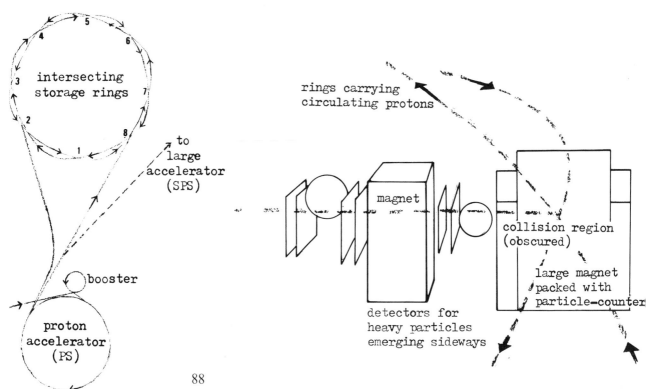

intersecting
storage rings

to
large
accelerator
(SPS)

booster

proton
accelerator
(PS)

rings carrying
circulating protons

magnet

collision region
(obscured)

large magnet
packed with
particle-counter

detectors for
heavy particles
emerging sideways

periment at CERN's
ersecting Storage Rings
R). Protons fed to the rings
m the accelerator will
ntinue circulating for hours
end. At eight intersection
ints protons travelling in
posite directions in the two
gs have the opportunity to
llide. One of the main
coveries with the ISR was
t heavy energetic particles
me out sideways from such
llisions – further evidence
small 'pips' in the protons.
e photograph and diagram
ow an experiment for
dying that process in detail
he Liverpool-MIT-Orsay-
utherford-Scandinavia
llaboration at Intersection 4.

ways that accorded well with the curious behaviour of quarks. One of the peculiarities was explored by David Politzer of Harvard University, among others. He described it thus:

'With other forces like gravity or electricity, the closer objects get, the stronger the interaction between them. But when you go closer and closer to a quark the colour force doesn't get stronger and stronger. On the contrary, when two quarks come close together they affect each other less and less. So, well inside the proton, the quarks are free to rattle around.'

The theory predicted that the quarks would enjoy short-range 'freedom'. But as soon as the quarks tried to move apart they felt a stronger and stronger force, suffering long-range 'slavery' which firmly held them in the proton.

This second peculiarity, the persistent strength of the colour force at a distance and the slavery that resulted, could be described in various ways. Kenneth Wilson of Cornell University offered a vivid picture of the quarks in a proton, tied together with strings.

Quarks on strings

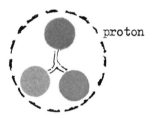

These were the fetters of the enslaved quarks. Wilson could draw more complicated pictures, with each string thickened into a bundle of many strings and the strings within the proton running in different ways. But the basic idea was simple and striking, once you were prepared to imagine gluons sitting still instead of flying about. Then the strings were made of gluons, in such a way that each gluon cancelled the colour of its neighbours, making the whole string white. The strings were very strong and would not stretch.

**Close-up view
of the string
– made of gluons**

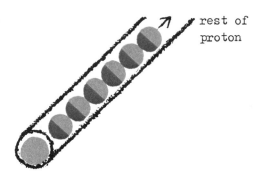

rest of
proton

If you imagined trying to pull a quark out of a proton, the only way you could move the quark was by making the string longer, by creating more gluons. (You might think of it as a chain rather than a string; you had to forge more links.) But the creation of gluons required energy, which did not diminish as the steps took the quark farther from its companions.

Suppose now, with great exertion, you managed to break the string. Would you not have liberated the quark? No, because that would be equivalent to asking for a piece of string with only one end. Snapping the string simply exposed new quarks at the broken ends.

**Breaking
the string
does not free
a quark**

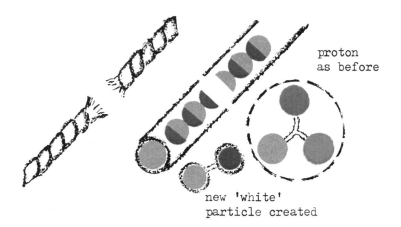

proton
as before

new 'white'
particle created

The exposed end of the string running back to the rest of the proton had the same colour as the original quark you were trying to remove. So you failed to change the proton. All you succeeded in doing, with all the effort, was to create a new particle. It was a perfectly conventional particle, too, consisting at its exposed ends of a quark and an anti-quark of opposite colours: in other words, an ordinary 'white' meson, or strong force-carrier.

There were rival theories about the precise nature of the force that enslaved the quarks: other kinds of 'strings' and theories which saw the

proton as a 'bag' with walls impenetrable to the quarks within it. The most important argument was about whether lone quarks could actually be set free, after all, and some of the theories allowed for that possibility. In more than ten years of trying, the experimenters had apparently failed to liberate a quark, yet they still might succeed with more powerful machines – in which case theories like Wilson's which made quark slavery absolute would eventually be consigned to the string recycling plant.

Indeed I need to re-emphasise that the colour force was just a theory, which had to be checked out by all possible experiments. What the experiments should be was none too obvious. But in the mid-1970s it was a very attractive theory, rich in its logic and its power to explain phenomena already known. It was apparently ideally suited to take its place in a self-consistent universe, as the description of the most powerful force of all.

Recalling the questions about quarks posed on page 82, the answers given by the central theory were then as follows. The glue holding the quarks together was coloured gluons acting on coloured quarks. The requirement for a gauge theory – that the gluons should affect one another – imposed severe restrictions on the possible games you play with colours and quarks. The simplest permitted theory coincided neatly with the analogy of mixing primary colours and anti-colours to make white particles. That description was a happy accident of no deep consequence: the quarks and gluons were not really coloured, but conformed to much the same sort of mathematical scheme as colours did.

And the scheme gave exactly the combinations of quarks found in the families of real particles: the proton and its relatives each composed of three quarks; the nuclear force-carrying particles each composed of one quark and one anti-quark. It predicted short-range freedom and long-range slavery and it offered an explanation of why nobody had seen a free quark.

The strong force deposed

As the theory of colour tightened its grip on their minds, physicists came to perceive the strong nuclear force as just a by-product of the superstrong colour force, its 'mother'. The binding force of the atomic nucleus, to which the Sun and the stars owed their energy, was feeble compared with the colour force between the quarks. The particles that carried that force, being white, neither exerted nor felt the full force of colour. But the physicists had met this situation before.

A similar relationship existed between other forces operating, right outside the atomic nucleus, among atoms and assemblies of atoms. When atoms combined to make molecules or rigid crystals, the electric force bound the atoms together with vigour. But after all the obvious opportunities for joining atoms together had been exhausted, and they had formed their molecules whether of two atoms or a thousand, a fainter force remained.

Molecules stuck together weakly by the Van der Waals force. It was named after the Dutchman who first defined the stickiness of molecules in the nineteenth century. The force became negligibly weak as soon as two molecules were more than a very short distance apart. Yet in very cold conditions even air and hydrogen gas would succumb to the Van der Waals force, when they turned to liquid. If the force managed to make a solid, it was generally soft, like candlewax.

A normal molecule was electrically neutral; a normal particle of nuclear matter was colour-neutral (white). The electric force contrived to create the Van der Waals force by setting the electrons in neighbouring molecules vibrating in sympathy with each other, even though they were tethered by electric force within their respective molecules. The colour force did much the same thing. Quarks were tethered by the colour force inside nuclear particles, but they could exchange relatively feeble 'vibrations' in the form of the white force-carrying particles.

By-product forces of these kinds might seem to be diminished in the cosmic scheme, but they were certainly not to be despised. The Van der Waals force helped to hold together ordinary objects on the Earth. The 'old' strong nuclear force was still the fire of the Sun. The colour force that gave rise to it was amply strong enough to create fierce conditions even when it was not really trying.

The restoration of a hierarchy

The quark theory and the rise of colour promised to rescue Western science from Oriental despair. The extravagant tricks that particles could get up to, with their 'borrowed' energy, had earlier seemed to some physicists impossible to reconcile with the traditional aims of scientific explanation.

Let me put the pessimistic argument in its extreme form, apologising in advance for the complications that are its essence. Particle X can be regarded, simultaneously, as a particle in its own right, a particle carrying a force between other particles, a particle composed of other particles, and a piece of a different particle.

For example the particle called a pion can leave a track in a bubble chamber, and is undoubtedly a particle in its own right. It is also a meson, a carrier of the strong force that holds the atomic nucleus together. However, a pion is capable of transforming itself momentarily into a proton and another particle (an anti-neutron, say) so that the proton is in a sense a piece of a pion. On the other hand, you can also make a proton from a pion and a neutron, so that the pion is a piece of a proton. The particles seem, on this view, to pick themselves up by their own bootstraps.

The apparent interchangeability of all particles and all roles led Geoffrey Chew from 1959 onwards to propound the theory known as 'bootstrap' or 'nuclear democracy'. It involved a conscious rejection of the traditional objectives of physics, to explain events in terms of forces acting between clearly defined particles. According to Chew none of the particles was more elementary or fundamental than any of the others. Nor should you enquire too closely into their detailed behaviour, or when and where they did their behaving. Instead you simply studied events – particles going into a reaction and particles coming out. The particles themselves ceased to be objects for analysis: they were more like conjectured connections between events.

If Chew were right, it might be very bad news indeed for Western philosophy and science, with their objectives of trying to dispel needless mystery from the universe by discovering its fundamental units and laws. A disciple of Chew, Fritjof Capra, revelled in the philosophical implications in *The Tao of Physics* (1975). In that book he claimed that, through bootstrap, the conclusions of the scientist converged with those of the Eastern mystic engrossed in his meditations. The mystic, preoccupied with the interconnections between all things, usually rejected the idea of atoms and fundamental units.

The dance of the particles, Capra said, became the Hindu's Dance of Shiva. And looking to the future he declared: 'Instead of a bootstrap *theory* of nature, it will become a bootstrap *vision* of nature, transcending the realms of thought and language; leading out of science and into the world of *acintya*, the unthinkable.'

The bootstrap theory (without these trimmings of unthinkability) was a refuge in the 1960s, when the known particles and forces made little sense. Physicists might console themselves by saying they were not meant to make sense. But by the 1970s Chew's theory was in eclipse. The rigmarole about the many guises of the pion lost its mystery once you accepted that pions and protons and neutrons were all composed of quarks. There was no way of making a quark from a pion, for instance.

And the 'undemocratic' hierarchy of particles was firmly restored, with quarks having special rank. As Abdus Salam put it: 'All particles are elementary but some are more elementary than others.'

Even when its own traditions were at stake, science would not decide the matter by taste, fashion or opinion polls of experts. The bootstrap theory made a very different prediction from the more fashionable gauge theories about what should happen at extremely high energies. The gauge theories would predict the existence of a limited number of types of quarks, from which nature could make a large but limited range of proton-like particles. In addition, there should be a variety of new force-carriers and perhaps some superheavy electrons. The bootstrap theory, on the other hand, implied unlimited possibilities for the creation of new forms of matter (or the old matter in new guises!) given ever-increasing supplies of energy.

Some day the question would be settled, perhaps by experiments with more powerful machines, perhaps by astronomical evidence about black holes or the extremely energetic Big Bang. Either Nature went on making novel particles indefinitely, or it did not. For the time being, the 'bootstrap vision' looked not so much wrong as irrelevant.

You could, if you wished, speak of some residual equality and interchangeability among the quarks and electrons themselves, but it was the equality of an inner cabinet. There was a clear hierarchy in behaviour, too, with the colour force operating exclusively on quarks, the strong nuclear force only on quark-composites, and the weak force and gravity operating on everything.

Hints of a possible return to a measure of democracy began to crop up in 1976. Theorists saw new ways in which durable particles of matter might be thought of as being generated by particles of the force-carrying type. But it seemed unlikely to restore the thoroughgoing equality of particles imagined by Chew and his associates.

A comparison with chemistry is not unfair. Before the revival of the atomic theory in the nineteenth century, it had been easy to form the impression that any kind of material might be changed into any other. The terrestrial scene and the alchemists' laboratories were full of examples of amazing transmutations. All of them simply involved changes in chemical compounds, but the misapprehension drove intelligent men to penury and madness in their efforts to make gold, before the stern hierarchy of elements versus compounds showed why their efforts were futile.

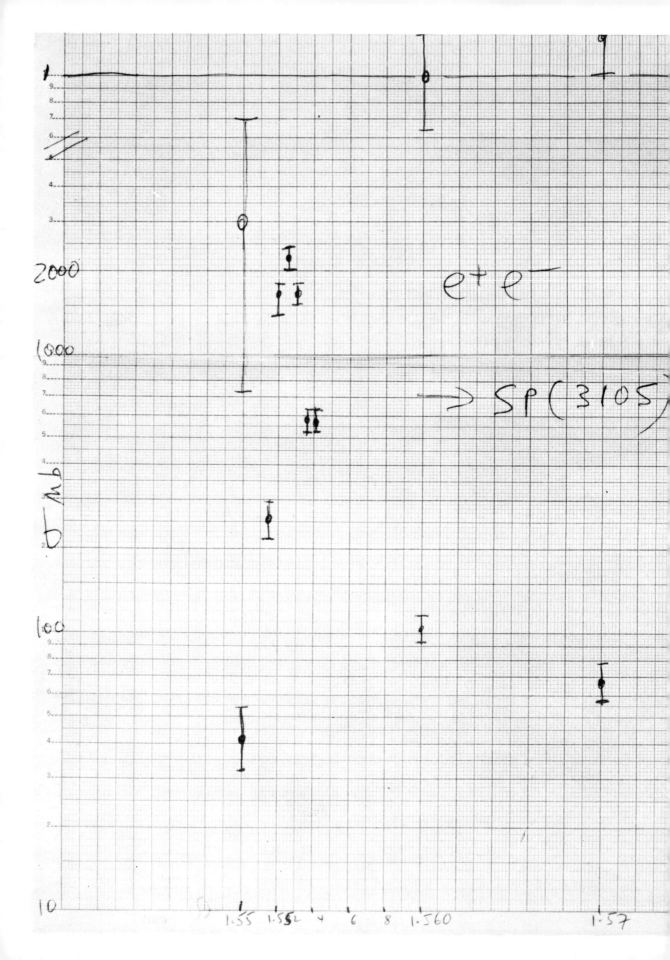

4 Charm

Physicists sometimes admit to a little fantasy that goes like this. One day they have a bright idea for an experiment. They write a proposal, gather their friends and their funds and build complicated equipment. At last all is ready. They lead the pulses of energetic particles into the midst of their apparatus and peer anxiously at their instruments. At that moment God convenes a conference of angels.

'Seen what those people are doing now?' He demands.

'Hey!' the chief angel says, looking down. 'They're crashing those dinky bits with 3 GeV at the centre of mass. And they're watching hard this time.'

'Well don't look at me,' says the Lord. 'It never really came up before.'

In the thoughtful pause that follows, one of the angels pipes up impatiently: 'We've got a million collisions already, and I don't know what to do with them.'

'We'd better take a vote then,' God declares. 'All those in favour of granting them a new particle please raise a wing. . . .'

Back on Earth, the counters start registering the result of the vote.

The repeated confirmation of human ideas in experiments was always thought-provoking. A few people scoffed saying that any inhabitable universe must have laws that intelligent beings could discover. For most practitioners there was a sharper sense of awe whenever ideas, especially of the crazier kind, were borne out by the readings of instruments. The speed with which it often happened, from the wish to the discovery, gave the illusion that nature had waited to behave that way until importunate human beings demanded it.

Charm was a discovery that came when it was needed. The idea of charm, as a new quality of matter, first cropped up in 1964 but few people took much notice of it. After the quark theory appeared, hundreds of ways of extending it were being reconnoitred in learned speculations. A decade later charm had become the obsessive quest for many of the world's physicists. By then it promised to crown a series of major advances in understanding the composition of matter and the cosmic forces at work within it. The first experimental hints of charm came in 1974, but they were awkwardly ambiguous. In 1975 clearer signs appeared. In 1976

[margin caption:]
moment of discovery. On [No]vember 1974 one of the [exp]erimenters working with the [SP]EAR ring at the Stanford [Lin]ear Accelerator Center [ske]tched this graph. In the [ma]chine, electrons and anti-[ele]ctrons of equal energy [(bo]ttom scale) were [an]nihilating each other and [cr]eating a new particle, which [wa]s then breaking up again [an]d throwing out new pairs of [ele]ctrons and anti-electrons [(ve]rtical scale). The outpouring [of] newly made electron/anti-[ele]ctron pairs rose to a sharp [pe]ak when the annihilating [ele]ctrons each had an energy of [jus]t over 1552 mass-energy [un]its (MeV), indicating a mass [of] 3105 units for the new [pa]rticle. (The circle with the [lon]g bars to the left of the peak [is] extraneous, being part of [an]other sketch-graph.)

97

the angels voted decisively and the counters registered satisfying evidence of 'naked charm'.

So what was charm? The basic point can be stated very simply. Until 1974 all known particles of heavy matter could be made out of just three types of quarks and their anti-quarks.

Charm was embodied in a fourth quark and its anti-quark.

The charmed quark implied a range of other particles, a whole new domain of matter, which could be made by this addition to a very select little band of basic particles. It multiplied nature's capacity to make proton-like particles, from combinations of any three quarks; also force-carriers for the strong nuclear force, from combinations of one quark and one anti-quark. For example:

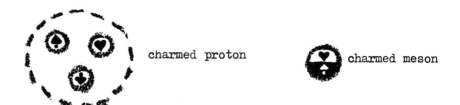

Upness, downness and strangeness, and later charm, represented different qualities of matter, akin to electric charge. The physicists came to call them quark 'flavours', but to avoid confusion I write only of quark types and qualities. The different types of quarks had different weights: the strange quark being heavier than the other two it always tended to give off energy and change into a lighter quark. The charmed quark was heavier still.

The qualities had no effect on the colour force between the quarks, described in the last chapter, which was indifferent as to type. They governed instead the possible transformations of one form of matter into another (or 'flavour dynamics', which illustrates the perplexing jargon I am trying to avoid).

Quark qualities were therefore very much wrapped up with the activities of the weak force as the Cosmic Alchemist. And it was in the study of the weak force, especially in the ultimately successful effort to unify it with the electric force, that charm seemed first possible, and then essential. It therefore neatly drew together, as it was required to do, the ideas about the weak force (Chapter 2) and the idea of quarks (Chapter 3). But some raggedness remained, and in this chapter the story of charm ends with the speculations about other quarks to come.

Another quark needed

'Charm, in the sense of a magical device to avert evil,' Sheldon Lee Glashow explained when I queried his use of the word. James Bjorken and 'Shelley' Glashow had first adopted the term in 1964 as a simpler expression of pleasure at a delightful idea. The basic constituents of matter apparent at that time were the three types of quarks postulated by Murray Gell-Mann and George Zweig and the four members of the electron family, the ordinary electron, the heavy electron and two distinct types of neutrinos. The second neutrino had only recently been discovered. At the simplest aesthetic level, it would be nice if there were a match, symmetrically four to four, between the two families, quarks and electrons. Similar ideas about a fourth type of quark were broached in Europe and Japan at the same time, though without Bjorken and Glashow's catchy tag of charm.

As the years passed charm took on a more serious aspect. It became the much-needed spell that might save an edifice of theory from collapse. In collaborations with his colleagues at Harvard University Glashow pursued the concept with special doggedness; he came to be regarded as the father of charm.

The evil that charm had to avert became apparent in the late 1960s. It was a failure of the theory that offered to unite the weak force with the electric force. The theory predicted, besides the conventional force-carriers for the weak force, W^+ and W^-, the new, uncharged force-carrying particle W^0. If it existed, W^0 ought to assist in the break-up of particles incorporating strange quarks. It would give recognisable products, different from those appearing during a break-up via W^+ or W^-.

Experimenters looked for appropriate sequences of events and failed to find them. For example, one strange force-carrying particle should have broken up like this:

eldon Lee Glashow,
rsistent predictor of the
armed quark. He was born
1932 and, after attending
h school in the Bronx, he
aduated from Cornell and
came a graduate student at
rvard. Glashow then
rked at several of the
rld's leading centres of
ysics – Copenhagen, CERN,
sadena, Stanford and
rkeley – before returning in
56 to Harvard, where he is
ofessor of physics.

99

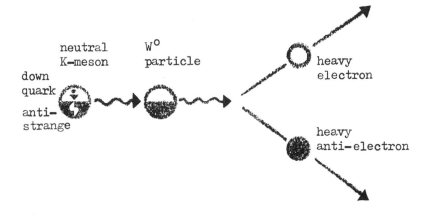

It did not happen: more precisely it was at least ten million times rarer than a very similar break-up of a strange particle, involving the conventional W$^+$ particle. So physicists began shaking their heads about the W^0, and doubting whether the theory that predicted it could be right. Later, the Gargamelle Collaboration found impressive evidence for the new form of the weak force, in other experiments. It seemed that the W^0 was real, so the mystery deepened about its inability to abolish strangeness.

By invoking charm, Glashow and his colleagues found they could explain why the strange particles did not use the W^0 as an aid in breaking up. In the old scheme, without charm, the up and down quarks were closely related and strangeness was the odd man out. With charm the strange quark had its own partner and, as a result, it became equal in status with the down quark. The up and charmed quarks were both supposed to have an electric charge of $+\frac{2}{3}$, while the strange and down quarks' charge was $-\frac{1}{3}$.

**Old scheme
of quarks**

**New
scheme**

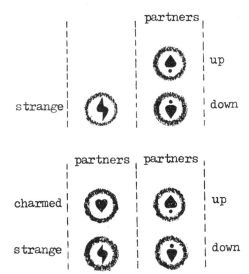

100

In the old scheme one could visualise a W^0 particle as consisting, say, of a strange quark and an anti-down quark, which could therefore annihilate an anti-strange quark. But that depended on the strange quark having a bachelor status. When all the quarks had equal status, in the new scheme, it meant that a strange quark would always be shadowed by an anti-strange quark in a W^0 particle, preventing the abolition of strangeness.

**Old view
of W^0 particle**

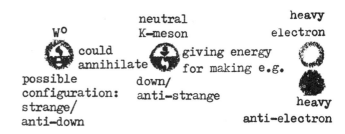

**New view
of W^0 particle**

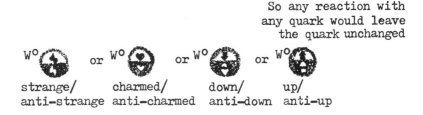

The prediction was of the existence of a new type of quark, representing a new quality of matter never seen before: charm. There was reason to expect that the charmed quark would be a good deal heavier than the three quarks already known. Nevertheless, it ought to be able to combine with other quarks to make many novel particles. So when news came of a novel particle in November 1974, it was natural to wonder whether it had anything to do with charm.

Caution at Brookhaven

A very odd wager was made at the Brookhaven National Laboratory on 22 October 1974. Samuel Ting bet Melvin Schwartz ten dollars that there was 'no resonance around 3 GeV'; in other words that there was no new particle of about 3000 mass-energy units. Ting at once wrote a memo to himself: 'I owe M. Schwartz $10'.

Modest bets between scientists on the outcome of experiments were common enough. What was unusual about Ting's bet was that he made it knowing he would have to pay up. He was trying to put Schwartz off the scent, because Schwartz hailed from Stanford, where another machine was fully capable of finding anything waiting to be found at around 3000 mass-energy units. There really was a new particle and Ting knew it well enough because he had discovered it at 3100 units.

Moreover it was a very peculiar particle. That it was more than three times heavier than the proton was not, in itself, surprising. Other particles of that sort of weight had been found before. Because they were bursting with mass-energy they were all very short-lived, breaking up just as rapidly as light could travel from one side of the particle to the other. The amazing thing about Ting's particle was that it survived a thousand times longer than that. By human standards, it was gone in a flash, but in the nuclear time-scale it was a veritable Methuselah. It therefore seemed to require a new law of nature guaranteeing its existence, and to represent a quality of matter never seen before – a new ingredient for building the universe.

But Ting was a cautious man, hesitating to publish until he was completely satisfied that his instruments were not hoaxing him. The upshot was that he had to share the discovery. The unhappy Ting was in a hotel in Stanford on Sunday 10 November, on the very day that Schwartz's colleagues found the particle at 3100 mass-energy units.

On 2 December 1974 the journal *Physical Review Letters* published three short papers. One by Ting and his colleagues from the Massachusetts Institute of Technology announced the discovery at Brookhaven of a new particle, called *J*. Another, by Burton Richter, Gerson Goldhaber and their joint team from Stanford and the Lawrence Berkeley Laboratory, told of the discovery of the same particle, named *psi*. The third paper was from Frascati in Italy. There Giorgio Bellettini's group in the ADONE laboratory had confirmed the existence of the particle (at the very limit of their machine's energy-range) within two days of hearing the news from the USA.

Inevitably, perhaps, there were dark hints that Stanford found the particle because they had heard what Sam Ting was doing; also counter-complaints about Ting's secretiveness. The Stanford team had a thorough explanation of why, for quite independent reasons, they happened to be looking at the creation energy of the new particle on that November weekend. And Burton Richter recalled his meeting with Ting.

'I met Sam at 8 o'clock that Monday morning and he said to me, "Burt, I have some interesting physics to tell you about." I said, "Sam, I have some interesting physics to tell *you* about!" While this is not sparkling dialogue, it began an astonishing conversation, as we had no idea about Ting's results.'

The substance of the dual discovery, as it unfolded at Brookhaven and Stanford, is more interesting than any recriminations.

he discovery of the *J* particle
Brookhaven. The graph
low shows the counts of
irs of electrons and anti-
ectrons at different combined
ergies, with a clear peak
ound 3100 mass-energy units
1 GeV). The dark portion
the graph shows results in
ugust 1974, and the light
rtion results in October
74, with a different setting
the instrument. The
ectrons and anti-electrons
re registered in the counting
om (below right) as signals
om the double particle-
tection system (right). The
agnets served to deflect the
ectrons upwards a little – out
the spray of miscellaneous
rticles emerging from the
rget. The instrument, with
o arms 18 metres long, was
nbedded in nearly 10,000
ns of steel and concrete when
e experiment was running, to
ntain the intense radiation
oduced by protons smashing
to the target.

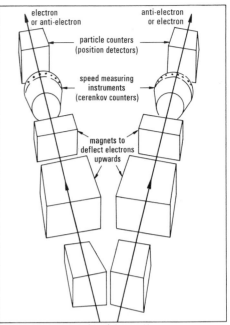

target (out of photograph)
proton beam from accelerator

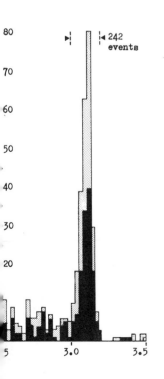

103

Samuel Ting, of the Massachusetts Institute of Technology, leader of the group that discovered the *J* particle at Brookhaven. He grew up in China but, having been born prematurely when his parents were visiting the USA in 1936, Ting was able to take up his American citizenship as a student at the University of Michigan. He has done much of his research in Europe, at CERN and at DESY in Hamburg. The *J* discovery made him a Nobel prizewinner in 1976.

In the 1960s the Brookhaven accelerator – sited on Long Island, not far from New York City – had been the pacesetter in high-energy research. The second type of neutrino and the omega-minus, the proton's strangest cousin, were two of its historic discoveries. But new machines overtook Brookhaven's in size and energy and everyone expected the machine, like an old lady, to go into a genteel decline. Instead, it pranced into a new realm of the micro-universe.

Ting was doing an experiment that many physicists thought a waste of time. For a start it was very tricky. Ting and his colleagues had spent ten years refining their techniques, and three years building their equipment. They were using the impact of Brookhaven's beam of protons on a metal target, just as a source of energy for creating new matter. Ting had enjoyed a close association with German experimenters at CERN and Hamburg and the team working with him at Brookhaven in 1974 included German members.

Out of billions of particles pouring through their equipment with every burst from the accelerator, they wanted to pick out just one electron and one anti-electron appearing in two big detectors at the same instant. Two elaborate detection systems, each with 5000 counters large and small, waited to trap, identify and time the flight of the electrons and anti-electrons. Pairs of electrons and anti-electrons, coming in sufficient abundance, would signal the creation and break-up of new kinds of matter. But other experimenters had passed that way before, and found nothing. Neither did Ting, for the first few months.

The detection system was originally set to a high range of possible creation energy. At the end of August 1974, it was tuned to a lower energy range and at once the detectors began registering the much-wanted pairs of electrons and anti-electrons. Dozens of them began showing up, all with the same combined energy for the electrons, close to 3100 units (3.1 GeV). It spelled the repeated production of a new particle of just that mass-energy, which was breaking down quickly, but not very quickly, into electron and anti-electron. A very short-lived particle would have too little time to establish a precise mass-energy, so that a succession of them would appear with rather different amounts. But the mass-energy of the particles of the new kind was precisely defined – a sign of their longevity.

Ting kept quiet about it, and insisted on checking and rechecking for months to make sure the particle was not an illusion created by faulty techniques. By the time he was satisfied, the research on the other side of the United States was coming to a climax too. The experiment in progress at Stanford was, in effect, Ting's in reverse: putting an electron and anti-electron together in the hope of building a new particle.

Champagne at Stanford

A ring called SPEAR lay at the end of the great three-kilometre electron machine of the Stanford Linear Accelerator Center (SLAC). That accelerator and its electrons contributed, as we saw, important evidence for the existence of quarks. But SPEAR was a special appendage to the accelerator, dedicated to the annihilation of electrons.

Just as the CERN laboratory at Geneva could store high-energy protons which circulated in opposite directions around the Intersecting Storage Rings, so it was possible to store electrons on the same principle, provided you kept on replenishing their energy. But there was an important difference: you could have electrons going one way and anti-electrons circling the other way. When they collided they would annihilate each other, producing pure energy from which new particles might form. The first big machine working on this principle was ADONE at Frascati. It created an abundance of heavy particles greater than anyone really expected. They were familiar particles, but their generous production was one of the early clues to forms of matter lurking in nature, beyond the known varieties.

The SPEAR machine handled more energetic electrons and anti-electrons. It began operation in 1973, with a maximum combined energy of 5000 units; by 1974 it was capable of 9000 units. Making the anti-electrons was quite easy. A third of the way down the two-mile accelerator, the beam of energetic electrons hit a target, creating anti-electrons, which were then accelerated for the rest of the way.

Many pulses of electrons and anti-electrons went into 'filling' the SPEAR ring. Thereafter the machine could work independently for hours, while a small minority of the electrons and anti-electrons collided. For Burton Richter, SPEAR was the fulfilment of a long ambition. A native New Yorker, he had come to Stanford eighteen years before, as a young experimentalist professing a love-affair with the electron: already Stanford was a prime centre for high-energy electron research. Within a few years he was caught up in the attempts, initiated by Gerard O'Neill from Princeton, to store electrons circulating in opposite directions in a ring. Pioneering a new technique of experimentation was frustrating and time-consuming, but by 1961 David Ritson and Richter were proposing to store anti-electrons and collide them with the electrons. As Richter was later to declare:

'I have been led on by a naïve picture: a positron and an electron, particle and anti-particle, annihilating and forming a state of simple quantum numbers and enormous energy density from which all the elementary particles could be born.'

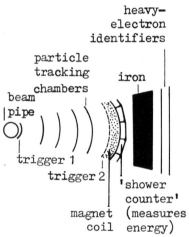

heavy-
electron
identifiers

particle
tracking
chambers   iron

beam
pipe

trigger 1

trigger 2

'shower
counter'
(measures
energy)

magnet
coil

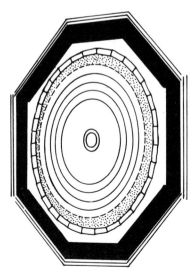

n exceptionally productive
ientific instrument. The
olenoidal magnetic detector'
SPEAR's West Pit
hotograph opposite) has
rved in a series of major
scoveries. As shown in the
oss-sectional diagram, it
nsists of concentric cylinders
detectors, designed to record
d track particles emerging
om the mutual annihilation
electrons and anti-electrons
the 'beam pipe'. The
mputer display (below)
ows an event recorded by the
etection system. A heavy
rsion of the 'gipsy' particle is
rowing out two pions
urved 'prongs') to make the
ormal gipsy, which in turn
reaks down into a very
nergetic electron and anti-
ectron (tracks almost straight
opposite directions). The
xperimenters at Stanford were
mused to see that the
esulting picture looked like
e Greek letter psi (ψ) – their
ame for the gipsy.

But there were long delays. Another man might have lost heart, and turned his energies elsewhere. Nine years elapsed before funds were forthcoming to build SPEAR, and thirteen years before Richter's 'naïve picture' was marvellously fulfilled. By 1974 he was 43, and in the midst of experiments with SPEAR.

When electron and anti-electron annihilated each other, they dissolved into pure energy very like a particle of light. But it was disreputable; it had no passport to travel through space-time, like a real particle of light. For a moment the bundle of energy remained in custody at the scene of the annihilation, until it transformed its energy into matter, creating pairs of particles and anti-particles that were permitted to travel.

Just how the energy materialised depended on how much of it there was. For instance, you had to 'freeze' 1876 units of energy to create the double mass of a proton and anti-proton. In theory, any particles could appear in any combination that fell within the budget of energy provided by the head-on collision of energetic electrons. Very little of the energy came from the annihilation itself.

By gradually changing the energy of the electrons and anti-electrons circulating in the SPEAR ring, the experimenters could tune their collision energy to the mass-energy of particular particles. If the energy were just right, those particles would form more readily than anything else. If they were heavy particles they would not survive for long. But instead of a jumble of miscellaneous matter the detectors would register signatures of those particles – in the form of the further particles that they produced as they broke up.

In 1974 the 35-strong experimental team from Berkeley and Stanford had surrounded one sector of the ring, in the so-called West Pit, with a very elaborate system of particle detectors. It recorded the products of the collisions within the ring. The experimenters had found inconsistent results between 3000 and 3200 mass-energy units. When they looked repeatedly at 3100 units, sometimes they had unusually intense bursts of particles in their detectors, but, more often, nothing special occurred.

To clear up this curiosity, the team arranged to control and measure the energy of the beam more accurately, so as to work their way in smaller steps through the region of inconsistencies. That was what they were up to, on that weekend in November 1974. As it was a more straightforward experiment than Ting's, the result was more dramatic and it came in hours rather than months.

When SPEAR entered the right energy range, the detectors began picking up an unmistakable barrage of activity, corresponding to the creation and break-up of a new particle. It became more and more

Burton Richter, mastermind of SPEAR. His 35-strong team from the Stanford Linear Accelerator Center and the Lawrence Berkeley Laboratory found the new particle called *psi*. It was the same as the *J* particle discovered at Brookhaven. Richter shared the 1976 Nobel Prize for physics with Samuel Ting.

intense, as the experimenters homed in on the exact energy. At 3105 units, the activity in the counters was seventy times stronger than usual.

On the Sunday afternoon, when the control room was crowded with people drinking champagne, Gerson Goldhaber of Berkeley drafted the paper announcing their discovery, and Richter revised it. They did not know that Samuel Ting, visiting from Brookhaven and alerted by the excitement in the Stanford air, was that very day phoning high-energy laboratories overseas, to tell them of *his* discovery.

It was a good day for the telephone company. The Stanford experimenters too were calling up their friends in other laboratories or, because it was Sunday, at home. As the word reached the theorists across the USA and around the world, they started scribbling down their ideas about the new particle. And what many of them were writing was *c* for charm.

The possibly charming gipsy

With Samuel Ting calling the new particle *J* and Burton Richter calling it *psi*, colleagues who were anxious to be fair began referring to it as the *J/psi*. After a while, 'gipsy' seemed a happy corruption and that is what I shall call it. But more important than its name was its constitution – especially whether it contained the charmed quark prophesied by Sheldon Glashow.

The discovery of the gipsy particle threw the world of high-energy research into creative turmoil. All the big laboratories capable of making gipsies and other possible particles of high mass-energy hurriedly revised their schedules. Research that had been fuzzy for a long time, with hundreds of able people following quite different hunches about the significant questions to address to nature, clicked into focus overnight. The laboratories gave priority to experiments that were relevant to understanding the gipsy.

Experimental teams, anxious to safeguard their investments of time and effort in complicated apparatus, urgently discussed how it could be switched to exploring the new realm of matter. Within the first few days a dominant theory emerged and helped to guide their planning. It suggested that the gipsy was a combination of a quark possessing charm, with its anti-quark equipped with anti-charm.

Suspected constitution of the gipsy (J/psi) particle

 or

It was a simple theory with immediate implications. The quark and anti-quark naturally had opposite electric charges so the combination was a little like a hydrogen atom. As physicists had known for half a century, the electron and proton in a hydrogen atom could adopt a variety of well-defined 'states'. The effective distance between them could increase in fixed steps, representing increases in the energy of the atom. Similarly, there ought to be more energetic forms of the gipsy, requiring correspondingly more energy for their creation. The energy steps were calculable: for example there might well be a new version at around 3670 mass-energy units.

Imagine the joy when the first of the energised gipsies showed up almost at once! The Stanford-Berkeley experimenters, using SPEAR, capped their own discovery. Just eleven days after they found the gipsy itself, with 3105 units of mass-energy, they encountered a new outburst of activity in their detectors. It corresponded to a particle of 3695 units, satisfactorily close to the prediction mentioned. (Later the masses were corrected to slightly lower values.)

After that, elation and confidence eroded and the first agony set in. The charm/anti-charm theory of the gipsy clearly pointed to several possible 'states' of this combination of quarks, characterised by different mass-energies. They were in principle convertible, one into another, by the intake or emission of energy. Experimenters knew roughly where to look, but months passed and the laboratories were silent.

The Germans in Hamburg were put on their mettle more than anyone else. There, the DESY (Deutsches Elektronen Synchrotron) laboratory had a newly completed ring system for storing energetic electrons and anti-electrons. It was an electron annihilator similar in principle to SPEAR and ideal, therefore, for following up the gipsy discovery.

In the German machine, called DORIS, electrons and anti-electrons circulated in opposite directions in separate rings and confronted each other at two crossover points. The places where electrons and anti-electrons could collide were marked by elaborate systems of detectors. One was called DASP and the other PLUTO, and at the time of the gipsy discovery two different experimental groups were using them.

They quickly agreed to set aside their intended research and pool their efforts in hunting for variants of the gipsy. Experimenters from Munich, Heidelberg and Tokyo were involved, as well as staff physicists of the Hamburg laboratory. They identified the two SPEAR particles without difficulty, but then there were problems. Despite meticulous engineering that made SPEAR look like an improvisation, DORIS was a new machine, not yet operating at full intensity.

109

SPEAR, the electron annihilator of the Stanford Linear Accelerator Center. 'Stored' beams of energetic electrons and anti-electrons, guided by electromagnets, circulate in opposite directions in a vacuum pipe. When they collide head-on, pairs of electrons and anti-electrons can annihilate each other. Their former energy of motion then becomes available for the creation of new particles. One advantage of the technique is its 'cleanliness': the annihilation of the electrons supplying the energy allows the products to show themselves without confusion.

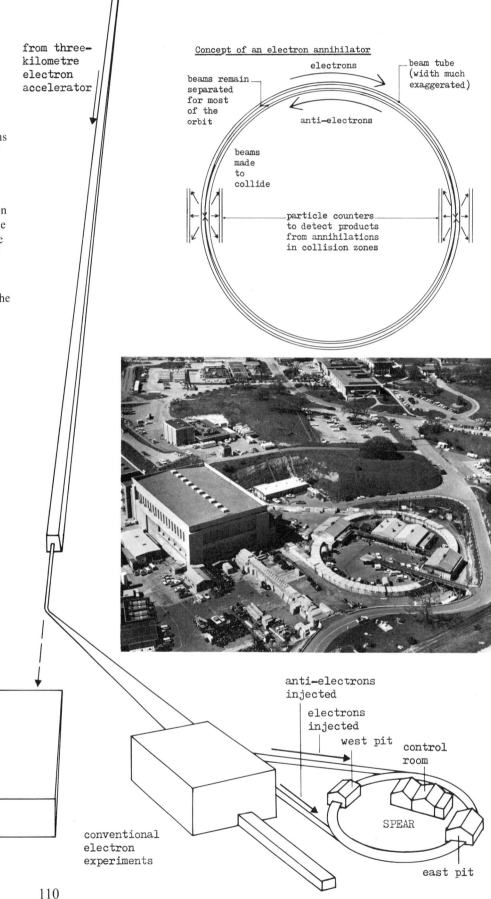

from three-kilometre electron accelerator

Concept of an electron annihilator

electrons

beam tube (width much exaggerated)

beams remain separated for most of the orbit

anti-electrons

beams made to collide

particle counters to detect products from annihilations in collision zones

anti-electrons injected

electrons injected

west pit

control room

conventional electron experiments

SPEAR

east pit

110

Week after week, the experimenters searched for evidence of further gipsy-like particles, amid their meagre harvest of electron annihilations. Theorists waited anxiously for further confirmation, from DORIS or SPEAR, of the various versions of the gipsy; also for other tokens of charm from other laboratories. After six months without news, many were already despairing.

At last the DORIS experimenters found what they were looking for. The heavier (3695) version of the gipsy already known gave out energetic particles of light (gamma-rays) at two different but fixed energies, to finish up as the lighter gipsy: That meant the particle was stepping down the energy scale, by way of a very short-lived intermediate version. The SPEAR experimenters came upon two other intermediate 'states' of the gipsy.

These results from DORIS and SPEAR, announced in July 1975, were reassuring. DORIS also gave hints of the original gipsy itself changing into a lighter form, at about 2800 mass-energy units. (It could be interpreted as representing a switch in the direction in which one of the quarks in the composite particle was spinning.) SPEAR, on the other hand, recorded renewed and complicated activity at around 4000 mass-energy units, part of which might well be due to still heavier versions of the gipsy.

The first agony of charm had abated, but fresh worries set in. One reason was that the charm/anti-charm theory of the gipsy was not exclusive or definitive. Any system of two charged particles might give a similar pattern of gipsy variants. Interpretations having nothing to do with charm could not be ruled out. For instance, Gordon Feldman of Johns Hopkins University and Paul Matthews of Imperial College, London, neatly fitted the conspicuous members of the gipsy family to the assumption that they were made of other quarks that were disclosing their colours in ways that the usual theories assumed to be forbidden.

Nor could the gipsy itself help in settling the issue in favour of charm. Supposing that the new particle did indeed consist of the charm/anti-charm combination, the charm was thoroughly hidden because it was self-cancelling. With zero net charm the gipsy could not be expected to show direct signs of charmed behaviour. It was reminiscent of a dubious Freudian argument that people who denied having sinister thoughts were repressing them.

But if the charmed quark were real, it should form combinations with the uncharmed quarks (up, down or strange) to create large families of particles in which the charmed quarks were not cancelled: particles with naked charm. Some of the possibilities are shown on the following pages.

111

Charmed particles predicted. The presumed existence of a heavy, charmed quark implied large numbers of novel particles incorporating it. There would be charmed relatives of the proton (combinations of three quarks) and force-carriers (combinations of quark and anti-quark), of which the latter would be less massive and easier to make. The gipsy itself, as a charm/anti-charm combination, has no net charm and so lies among the uncharmed particles known earlier. (Compare this diagram with the one on pages 76–7.)

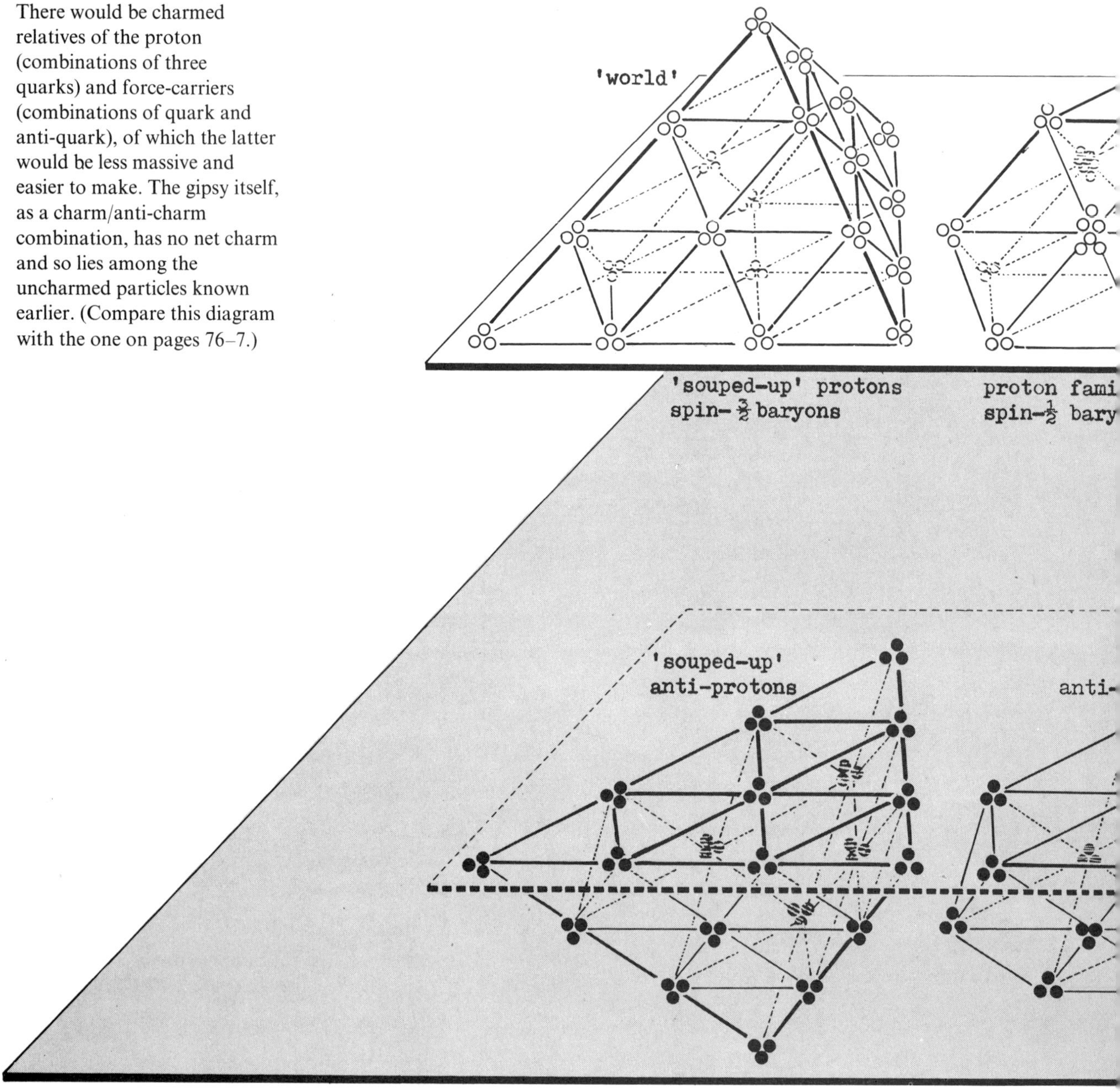

'world'

'souped-up' protons
spin-$\frac{3}{2}$ baryons

proton fami
spin-$\frac{1}{2}$ bary

'souped-up'
anti-protons

anti-

112

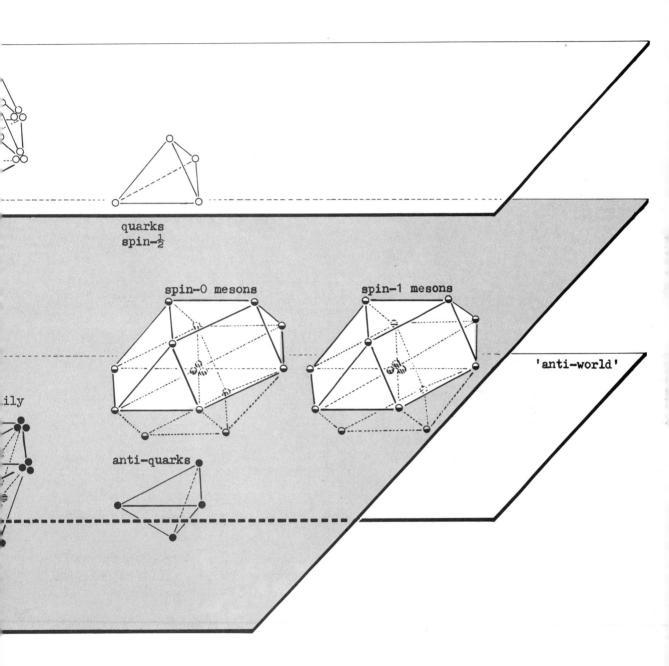

quarks
spin-$\frac{1}{2}$

spin-0 mesons

spin-1 mesons

'anti-world'

ily

anti-quarks

The 'flat' patterns of particle families made with combinations of the first three quarks became elaborate, three-dimensional patterns when you mixed in the charmed quark.

In an imperatively titled paper *Search for Charm* published in April 1975, three theorists set out an agenda for the experimenters. Mary Gaillard, Benjamin Lee and Jonathan Rosner had written most of it before the discovery of the gipsy, and anticipated the appearance of a charm/anti-charm particle. But they also enumerated many possible reactions in which charmed particles could be created and detected. They suggested that one or two experiments might have 'seen' charm already, but the general sense of their paper was that there were plenty of ways of finding charm.

The second agony of charm lasted from July 1975 to May 1976. The trouble was that experiments that should have come up with quick and convincing evidence for some of those particles with naked charm failed to do so, and failed repeatedly. Encouragement for the 'believers' was forthcoming in a few rare bubble-chamber pictures, yet for sceptics that only deepened the mystery of why experiments using particle-counters gave no very compelling sign of charm. The search for it became almost exasperating.

The marks of charm

How did you recognise charm when you saw it? The clue came from the original purpose for which charm was invented by the theorists. It was to provide a partner for the strange quark, in the theory of the weak force. And that alchemical force could transmute the one into the other fairly readily. As the charmed quarks were bound to possess more mass-energy than the strange quarks, it would be natural for them to break down into strange quarks, accompanied by such other particles as were needed to keep the accounts of energy and electric charge in good order.

Charm in decline would change to strangeness. I call that the Garbo effect, alluding to the admired actress who became a recluse. So the question about ways of identifying charm shifted to the recognition of strangeness. The commonest strange particles consisted of one strange quark in combination with an up or a down quark and they usually broke down in a characteristic fashion into two or three non-strange particles.

To find such signs of strangeness appearing in circumstances where, in theory, particles with naked charm ought to be produced was a suitable task for bubble chambers (p. 41). Prompted by Gaillard, Lee and

114

Rosner's *Search for Charm*, the experimenters realised that neutrinos (neutral electrons) shooting through the liquid of a bubble chamber ought to be able to generate charm.

In the imagined sequence of events, an incoming neutrino first reacted by the weak force with the quarks in the nucleus of an atom in the bubble-chamber liquid. The neutrino changed into a heavy electron, while one of the quarks changed into a charmed quark. The resulting charmed particle would then break up quickly. In one distinctive process, a neutrino and an uncharged strange particle would appear. The strange particle would itself decay into two oppositely charged 'normal' particles. Of these various particles, the neutrinos and the uncharged strange particle could leave no track. Schematically, then, the events might look like this in the bubble chamber (the dotted lines are not tracks).

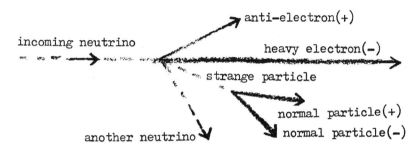

The strange particle, breaking into a tell-tale 'V', would be the fulfilment of the Garbo effect. The Gargamelle Collaboration, which had already discovered the Starbreaker, was by now enlarged to eighty physicists from 13 European centres. Many of the same individuals remained prominent in the enterprise, including Fred Bullock, Don Cundy, Ettore Fiorini, Pierre Musset, Don Perkins and André Rousset. They set themselves the task of trying to find such a pattern of events in their bubble-chamber pictures from CERN. So did a group of American and European experimenters using the big bubble chamber at Fermilab (Berkeley – CERN – Hawaii – Wisconsin Collaboration).

The Fermilab bubble chamber was made more suitable for neutrino experiments by adding heavy atoms (neon) to the liquid hydrogen with which it was originally filled. In addition, a detector of heavy electrons installed behind the bubble chamber helped to narrow down the possible interpretations of the particle tracks recorded in the bubble-chamber photographs.

As usual, the teams on both sides of the Atlantic had to search patiently through many thousands of photographs, looking for rare events. And, by the end of 1975, both had reported success. They found evidence for charm in several of the photographs matching exactly the conjunction of

115

A candidate for a charmed particle from a bubble chamber at the Brookhaven National Laboratory. The photograph was interpreted as showing the effect of an invisible neutrino (entering from bottom right) reacting with a proton to produce a heavy charmed particle. It was apparently a relative of the proton. The 'V' signature, produced by the decay of a strange particle that leaves no track itself, lies within the conspicuous loop formed by a normal charged particle curving under the influence of the bubble chamber's magnet.

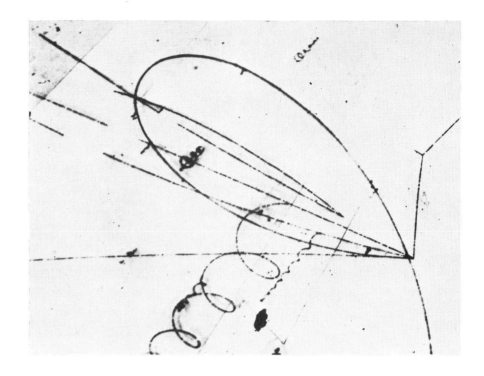

The first candidate for a charmed particle from the Gargamelle bubble chamber at CERN. The charmed particle itself is too short-lived to leave a track, but compare this photograph with the diagram on the previous page showing what might be expected from its break-up. The neutrino coming in from bottom left of the photograph leaves no track. The track labelled 1 is the anti-electron; 6 is the heavy electron. Tracks 2 and 3 are the 'normal' particles produced by the break-up of a strange particle which leaves no track, but the 'V' made by tracks 2 and 3 clearly points back to the scene of the reaction. Those tracks constitute the expected characteristic 'signature' of a charmed particle.

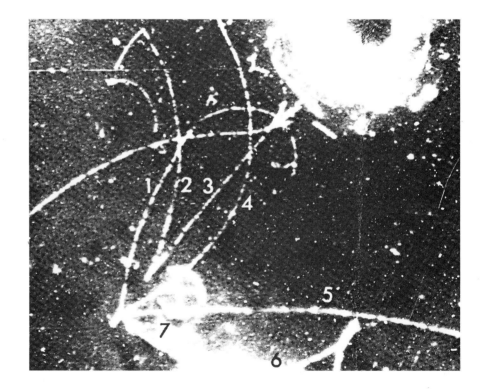

events predicted schematically for the formation and break-up of a charmed proton. One of the photographs appears opposite.

Earlier, in April 1975, the Brookhaven laboratory had published a single, very curious bubble-chamber picture, offered as evidence for charm. Here, a neutrino seemed to be charming a proton, changing it into a heavier particle that quickly broke into a strange particle (Garbo effect) and a swarm of 'normal' particles. By the interpretation of Nicholas Samios and his colleagues, the events involved a doubly-charged, singly-charmed relative of the proton.

The small harvest from the bubble chambers was encouraging for the theorists of charm. But it was an insufficient foundation to support a whole new scheme of matter in the universe. There were misgivings about some of the details. Some experts said that in the first batches of bubble-chamber pictures from Fermilab the strange particles left their signature too clearly too often. In any case, neutrinos were an extremely inefficient tool for making anything. If the inert neutrinos created charm, more reactive particles ought to do so abundantly.

From the viewpoint of the experimenters, the discovery of the gipsy was something that theory could not tarnish, but could certainly confound. Like Columbus they had crossed a virgin ocean and discovered a new continent of matter. Columbus, brainwashed by his own theory, thought he had reached Asia and called the aboriginal Americans 'Indians'. Was the legendary continent of charm like that?

Several leading experimenters voiced their growing scepticism about the supposed new quality of matter. If decisive evidence for naked charm were not forthcoming within a few months, or perhaps by the end of 1976 at the latest, the theorists would simply have to go away and think of something better than charm.

The interplay of theory and experiment always meant subtle strategies and wary assessments. An experiment that conflicted with a well-established theory was more likely than not to contain some error, yet a single well-attested result could destroy the most venerable of theories. Unexpected findings provoked new thinking, while ingenious theories like charm often stimulated experiments – but only for as long as they looked fruitful.

When we toured the major laboratories early in 1976, there were plenty of experiments, using particle-counters, busy trying to find charm. For example, one of the hunts for charmed particles, at the Intersecting Storage Rings at CERN, had been authorised three days after the discovery of the gipsy. Groups of American and German visitors worked with CERN colleagues and found that, as protons in the circulating

Massive particle detectors at Fermilab that figured in the search for charm.

beams slammed together head-on, they often produced, coincidentally, an electron and heavy electron of opposite electric charge.

That was meant to be a token of charm, because it could imply the simultaneous creation of a charmed particle and anti-particle, one of which threw out a heavy electron, say, and the other an anti-electron. The experimenters said they were seeing the right patterns of events in their particle-counters. Yet some colleagues criticised the experiment as somewhat 'unclean', saying there were too many ambiguities and uncertainties in it.

At the next intersection around the CERN rings, another experiment was about to look again at that production of electrons and heavy electrons of opposite charge. Burton Richter from Stanford, co-discoverer of the gipsy, was spending a year at CERN. Some French experimenters were already using elaborate detectors of ordinary electrons in a search for charm, and Richter had persuaded them to let him install a system for detecting heavy electrons.

Across the Atlantic, Fermilab too was in hot pursuit of charm. Indeed one of the big experiments had been gathering possible evidence for charm since long before the discovery of the gipsy: evidence for something, anyway. The particle detectors stood at the end of the kilometre-long bank of soil used to purify Fermilab's beam of high-energy neutrinos. A team led by David Cline of Wisconsin, Alfred Mann of Pennsylvania and Carlo Rubbia of Harvard placed a tank containing sixty tons of scintillating liquid in the path of the neutrinos. If a neutrino reacted and set loose any charged particles, the liquid would glow momentarily. Beyond was an array of 'spark chambers', strung with many electric wires in which small sparks would mark the passage of charged particles.

The last of the spark chambers were standing behind a massive wall of iron and massive magnets too; together they barred the passage of all charged particles except heavy electrons. About once in every thousand cases, when heavy electrons appeared and reached the last of the detectors, two of them arrived simultaneously.

Cline, Mann and Rubbia were sure that these rare double appearances of heavy electrons (dimuon events, as they called them) signalled the creation of a new form of matter. They spoke of Y particles, possibly including different types of particles, lying in the range of 2000–4000 mass-energy units. An obvious suspicion was that they included charmed particles. The experimenters were, though, the first to admit that they had too few results to rule out other possible explanations, and set to work to enlarge their detectors.

118

And then there was SPEAR, the ring at Stanford where the gipsy itself had shown up so clearly in November 1974. Routinely they were annihilating electrons and anti-electrons, creating concentrated pools of energy in the very ranges at which the particles with naked charm should be created. If the particles were there, why weren't they being detected?

After a year and a half, the theorists who favoured charm were far into a second period of worrying and waiting. I worried, too, in those first few months of 1976. This book and the associated television programme were intended to lay before the public the splendid fruits of recent research into the workings of the universe. But were the fruits after all rotten?

If charm were not real, then other ideas might begin to decay. The experimental facts remained blight-proof, of course. The reality of the newly discovered force (the Starbreaker) and of the gipsy itself as a new form of matter were unaffected. An absence of charm would nevertheless imply some profound misunderstanding about the nature of particles and the forces between them. If strangeness had no stranger companion, elation would give way to bewilderment.

Murray Gell-Mann had a general 'law' about the behaviour of particles. It said that anything not forbidden was compulsory. By this harsh test, particles with naked charm *had* to exist, wherever there was enough energy to create them. That 'wherever' certainly included several of the big machines operating at the time: Fermilab's monster, CERN's Intersecting Storage Rings, SPEAR at Stanford and DORIS at Hamburg. It seemed fitting that Gell-Mann himself should mention to me, during a phone conversation early in May 1976, that SPEAR was apparently 'seeing a charmed meson'.

Sighs of relief

From SPEAR, Stanford's electron annihilator, came the first evidence for charm that was naked enough and convincing enough to cause sighs of relief among the theorists. By scoring again in this way, SPEAR was proving to be an exceptionally productive piece of machinery. And it was the same team from Berkeley and Stanford, using the equipment in the West Pit, that served to find the gipsy. With Burton Richter away at CERN, Gerson Goldhaber of Berkeley was left as the senior experimenter concerned with charm.

Goldhaber had been a co-discoverer of the gipsy in 1974, in the course of a sustained collaboration using SPEAR involving teams of physicists from the Stanford Linear Accelerator Center and the Lawrence Berkeley laboratory. He was born in Germany in 1924 and as a Jewish refugee

Goldhaber learned his physics at the Hebrew University in Jerusalem. He eventually arrived at Berkeley in 1953 to work with the Bevatron, the most powerful accelerator then available. He took part in early research on anti-protons and the exploitation of the newly-invented bubble chamber, before the construction of the SPEAR ring and its quite different techniques lured him away.

Finding the expected charmed particles at SPEAR was largely a matter of working out why their presence was not altogether obvious. The answer was that the system of detectors in the West Pit was not well suited to recognising the typical signature of charmed particles. The particles would usually break up into many pieces, which swamped the equipment with indecipherable information. Only rarely would a charmed particle break up in a fashion tidy enough to signal its presence clearly.

Goldhaber and François Pierre, on leave from the French accelerator laboratory at Saclay, independently decided to go back through a year's records, looking harder for those rare events. That meant reviewing the outcome of more than twenty thousand electron annihilations, searching particularly for strange particles coming from a putative charmed particle of a particular mass-energy. They both found examples of what they were searching for; when they realised they were both after the same thing, they worked together and soon had a collection of about a hundred appearances of a new particle.

The 'appearances' were deduced, of course, rather than seen directly, but they pointed persistently to an uncharged particle of 1865 mass-energy units. Other members of the team joined Goldhaber and Pierre in uncovering further evidence of the new particle and further clues to its behaviour. It broke up in at least two different ways, both involving the wanted strangeness. Moreover, there were signs of another particle possibly a modified anti-particle of the first, being created at the same time.

The new particle was exactly what was ordered by the father of charm Sheldon Glashow and his colleagues Alvaro de Rújula and Howard Georgi. In their theoretical scheme (1975) the lightest sort of charmed particles ought to be about 1900 mass-energy units, and break down exactly as described. The prototype neutral particle in this group was a combination of a charmed quark and an anti-up quark. That, together with its equal and opposite particle (anti-charm plus up), was what SPEAR was apparently creating.

They were force-carrying particles (mesons) and the next landmark would be the discovery of a charmed proton in a particle-counter experiment. The same theorists, Glashow, de Rújula and Georgi, were circulating in 1976 an estimate that the lightest relative of the proton

Charmed particles at SPEAR, 1976

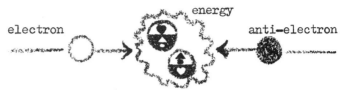

The particles created from the energy of the electron annihilation include pairs of charmed and up quarks.

The charmed/anti-charmed pair could just constitute a gipsy particle, and the up/anti-up an ordinary meson. But, if they group differently, the result is two particles with naked charm.

They can vary in this respect: the direction of spin of the two quarks in the particle can be the same or opposite.

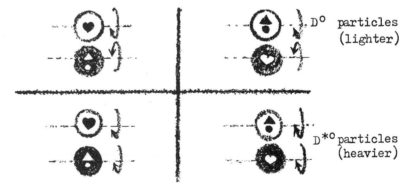

The interpretation of the break-up of the particles observed at SPEAR was that, in each event, one of the D^0 particles, of 1865 mass-energy units, and the oppositely constituted D^{*0} particle, of 2020(?) units, were formed at the same time.

with any charm would have a mass of 2260 units. The particles in question would be a proton in which an up quark had been replaced by a charmed quark.

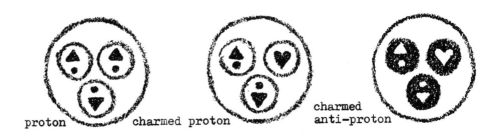

proton   charmed proton   charmed anti-proton

The charmed proton turned up in the form of its anti-particle, an equally valid representative, in Experiment 87 at Fermilab. Wonyong Lee and his colleagues in a Columbia-Fermilab-Hawaii-Illinois collaboration announced its discovery in August 1976. They made the charmed anti-proton by shooting very energetic particles of light (gamma-rays) at a target of beryllium.

In their detectors they looked for a characteristic signature of the expected particle – a pattern of other particles produced by its rapid break-up. They wanted cases where three normal force carriers (two negative pions and one positive pion) appeared at once, together with the obligatory strange particle. The last would be a strange neutral relative of the proton, which in turn broke up after a short delay in another characteristic fashion – into an anti-proton and a positive pion. In short they needed a combination of five particles showing up in the detectors.

After recording 15 million events in their target, and eliminating some ambiguous results, Lee's team reported that they had found 50 cases of the production and break-up of the charmed anti-proton. They deduced that the mass of the new particle was between 2250 and 2270 mass-energy units – in astonishingly close agreement with the figure of 2260 predicted by the theorists. The break-up was also in full accord with the theory.

By this time the remaining sceptics were in retreat. The families of new particles were showing up with exactly the features predicted by the charm theory, and the physicists could settle down to the more routine task of identifying all the members of the various families. While the details of each experiment and its interpretation could always be questioned, there could no longer be much serious doubt about the existence of new objects in nature, fitting the theory. There remained only the need to prove a subtle point about the charm theory, namely that it would allow one charmed particle in about twenty to break up *without* changing into a strange particle.

Of course, the people who had been claiming evidence for charm earlier, especially in the bubble-chamber experiments at Brookhaven, CERN and Fermilab, could now say 'we told you so'. In a longer perspective of history they will perhaps win the credit they deserve. The alarm about missing charm in the experiments using particle-counters really came about because the counter systems were barely adequate for the task. Why should the 'angels' vote for it if there was no compelling need?

Remember that only eighteen months elapsed between the announcements of the gipsy and the SPEAR charmer. To the theorists it seemed like a long time. But it was a short while in the timescales of high-energy experiments, which involved building complicated arrays of detectors. The particle-counter experiments that looked for naked charm – including the successful one at SPEAR – were hasty improvisations for a very tricky task. When other experimenters claimed they were seeing charm they were probably right, but there were too many loopholes in the experimental systems and in the interpretations of the results for their claims to carry sufficient conviction.

The bubble chambers had the advantage that they put on record whatever games nature chose to play. They were more versatile. Particular kinds of events might be very rare but if the suggestion of a new reaction came up you could scan the films, looking for it – and look again and again if need be. A properly designed counter experiment could give results in days or even minutes, but tortoise-like research with bubble chambers could win the race.

In *Search for Charm*, Gaillard, Lee and Rosner had written: 'The most convincing evidence for charmed particles would come from observation of short tracks.' In other words, it would be nice to have a *picture* of one of the charmed particles. They were too short-lived to leave any discernable trace in a bubble chamber. There remained the possibility of using photographic emulsion. Many 'stacks', each consisting of hundreds of layers of fine-grain photographic material, were placed in the path of high-energy particles. The hope was that a particle with naked charm might form and take its own photograph.

Photographers have learned to be wary of baggage checks with X-rays that can fog undeveloped film. But take photographic emulsion of sufficiently fine grain and deliberately expose it to energetic particles and you can find under a microscope the tracks of individual particles. This technique was thoroughly exploited in the study of cosmic rays, using stacks of emulsion flown high in the atmosphere on balloons. Applied to the search for charm, it might reveal a distinct track if a charmed particle

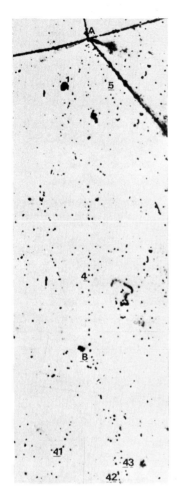

The first candidate track of a particle with charm was released before the end of 1976. The vertical dotted track of photographic grains has been greatly magnified because the particle travelled scarcely a fifth of a millimetre before it broke up, implying a lifetime of less than a million-millionth of a second. The track turned up when physicists in Brussels, Dublin, London, Rome and Strasbourg hunted through stacks of photographic emulsion that they had exposed to neutrinos at Fermilab. Although other interpretations could not be completely ruled out, charm was the most likely explanation for the pattern seen.

survived long enough only to travel a thousandth of a millimetre through the emulsion. *Search for Charm* predicted tracks of perhaps a tenth of a millimetre.

An enthusiast of long standing for the emulsion technique, Eric Burhop of University College London, organised one of the more elaborate and hopeful searches for charmed particles at Fermilab. A big international team arranged to stand half a dozen stacks of the special and very expensive emulsion in the neutrino beam for several weeks, while detectors tried to pinpoint the origin of particles emerging from the emulsion. That was intended to help in the subsequent 'needle-in-a-haystack' search of the emulsions, carried out by teams of men and women with microscopes at laboratories in six different countries. It was a tricky task, but they found the hoped-for tracks. The best among the early candidates turned up in the participating laboratory in Brussels. So, before the end of 1976, the year of charm, there was a picture of charm.

The caution about admitting charm was proper, and so was the persistence in trying to reconfirm it. It was an exceptionally important issue. For all the theoretical advances in understanding the weak force and its connections with other forces, charm was an indispensable keystone. With its discovery, the theory looked solid indeed.

As in the previous chapter, I now draw a metaphorical line across my narrative. Everything said up to this point fits together in an impressive monument built of theory and experiment in just proportions. It will be surprising if it does not stand for many years. But charm was a crazy idea in 1964, and a decade later other theoretical schemes and imperfectly understood experimental results were jostling for position in the new physics. There was no way of telling what would endure and what would vanish in the near future, or indeed whether the trends of thought were in the right direction.

Charmed universe

The theorists certainly wanted charm, but did nature itself really need it? Or was charm just a piece of decoration in the micro-universe, an extravagance that helped to keep physicists in employment? If you wanted to be tough-minded you could say that the universe was built of two kinds of quarks (up and down) plus the ordinary electron, and extras like strangeness and charm were of little practical significance.

Hydrogen was the prime raw material of the universe because its nucleus, the proton, was the lightest combination of three quarks (two up and one down) that the natural scheme permitted. Any other comparable particle containing heavier quarks quickly broke up to make protons. Add electrons in the protons and you had hydrogen gas, ready for making stars.

In the more elaborate atoms, fabricated in stars, all that really changed were the proportions of up and down quarks, and the numbers of electrons needed to complete the atoms. To these simple recipes you had to add some cosmic forces. Of these, the colour force that glued the quarks together in threes, to make protons and neutrons, knew nothing of strangeness or charm, because it was wholly indifferent to the types of quarks on which it acted. Similarly the electric force that held the electrons in place responded only to electric charge. The strong nuclear force that held the atomic nucleus together was carried mainly by force-carrying particles consisting of up and down quarks and anti-quarks.

At this point, though, the sceptical view of strangeness and charm began to break down. Strange particles certainly took part in the routine workings of the strong nuclear force and therefore helped in making the Sun and the stars burn and preserving the integrity of ordinary matter. Charmed particles would do likewise, though to a lesser extent. Ordinary matter possessed no net charm, because it was always cancelled precisely by anti-charm. But ordinary matter had no net electric charge either; the charges carried by the electrons and the atomic nucleus neutralised each other. That did not mean electric charge was unimportant.

Charm had to pervade the universe in fact. Nature was capable of creating any kinds of possible particles at any time, using the energy that could be 'borrowed' for a moment, thanks to the uncertainties about energy. The anti-particles had to be created at the same time, of course, so that the total matter of the universe remained unaltered. But in this respect charmed particles differed from more commonplace particles only in being heavier, requiring a larger borrowing of energy for the creation of charmed/anti-charmed pairs. Therefore they were somewhat shorter lived, but the difference was not very great.

For these reasons, charm would be present in all ordinary matter. At any instant, the human body for example would contain a quantity of charmed quarks. Although you could never catch them and weigh them, you might reasonably claim to have about an ounce of charm in your body – but an ounce of anti-charm as well. And there would be a small minority of atoms briefly possessing naked charm.

What difference would it make, if charm were not present in the universe? Perhaps nothing conspicuous: otherwise the need for charm would have become obvious sooner. But the workings of the universe, guaranteeing human existence, depended on many subtleties such as

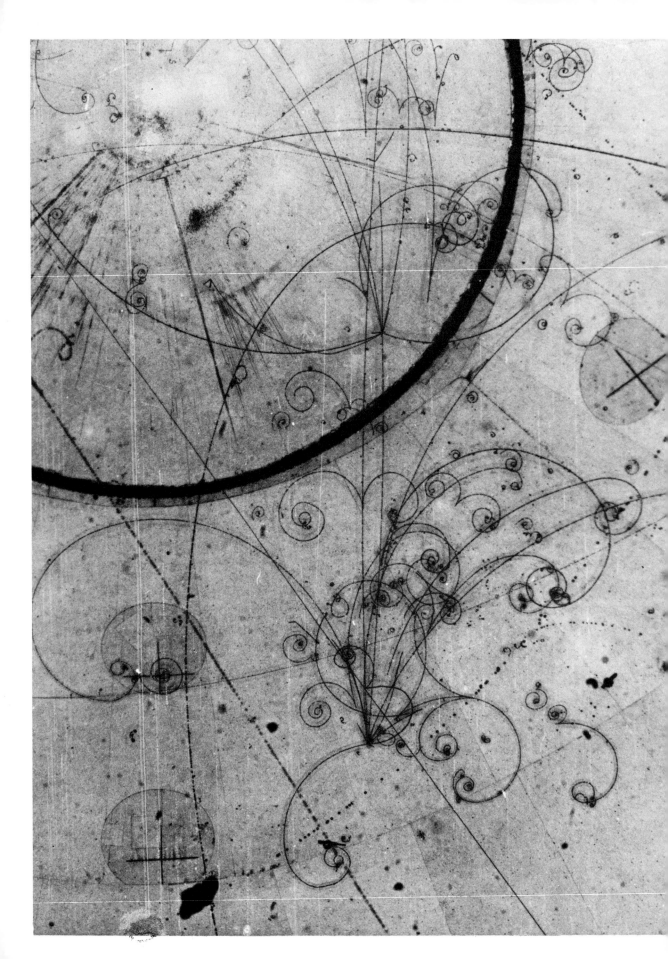

the new version of the weak force that I have called the Starbreaker. If charm did not exist, the behaviour of the strong nuclear force might be affected slightly. The rate at which stars burn depends critically on the relative strengths of the strong nuclear force and the electric force.

Charm must also have figured prominently in the Big Bang with which the universe began. When all matter and energy was concentrated in an extremely small volume, pairs of charmed and anti-charmed quarks would have been created in vast numbers. The number of different types of particles available to nature had a direct effect on the temperatures prevailing during the Big Bang: the more there were of them, to share the energy, the lower the temperature would be at any instant. The course of events during the first split second of the universe's existence presumably had a decisive effect on the subsequent overall character. Charm might, for example, have affected the creation of small irregularities in the early universe which eventually caused matter to concentrate in galaxies. But, after the discovery of charm, it remained for the theorists to calculate all its possible effects.

And charm was not necessarily the end of the story. The account of the weak force as Cosmic Alchemist, transforming one kind of particle into another, was a little ragged at the edges. A theory that dealt in trans-mutations ought to define very precisely what particles could be created, but it remained an open question, just how many types of quarks and electrons there were.

Still more particles?

Taking a simple view of the basic constituents of the universe, as known in the mid-1970s, you could guess they were eight in number: four types of quarks and four types of electrons. But did the families of quarks and electrons contain more members – new fundamental particles waiting to be discovered, beyond the charmed quark?

By 1975 there were indications that yet more forms of matter might exist. From SPEAR came reports of evidence for a superheavy electron. It made life awkward for the experimenters because it was created in the energy range already being scanned for proof of charm.

Martin Perl and his colleagues detected peculiar events occurring in SPEAR. From the scene of collision an electron and a heavy electron (the well-known muon) carrying opposite electric charges were ejected at the same moment without any other detectable particles coming out. No conventional process, involving conventional particles, could account for such events.

127

The experimenters offered this interpretation. They said that the energetic annihilation of electron and anti-electron in the colliding beams was producing a superheavy electron or 'heavy lepton' which they named *U*. Its anti-particle was produced simultaneously. The one broke down into an electron and the other into a muon. The decays would produce undetectable neutrinos as well.

The particle called *U* was grotesquely weighty for an electron. Theorists had been asking one another for many years why the muon, a heavy electron of 105 mass-energy units, was two hundred times heavier than the ordinary electron (0.5 units). They had no answer even to that. And the *U* particle was estimated to be 1800 mass-energy units, twice as heavy as a hydrogen atom!

Haim Harari of the Weizmann Institute in Israel advanced a scheme that commanded interest. He envisaged two additional pairs of particles, making a total of six quarks and six members of the electron family.

Extended quark family

Extended electron family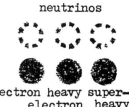

neutrinos

electron heavy super-
electron heavy
electron

The proposed additions to the electron family consisted of a superheavy electron ('heavy lepton') and a new neutrino to go with it. The extra quarks were those labelled *t* and *b*, which Harari intended should stand for 'top' and 'bottom', exactly their relationship when the pair was represented in that way. But some physicists, already disgruntled about 'up' and 'down' and finding more euphony in 'strangeness' and 'charm', began calling the supposed new qualities of 't' and 'b', 'truth' and 'beauty'.

The superheavy electron lent its weight to the argument for schemes such as Harari's: it was easy to suspect a natural symmetry that matched the number of quarks with the number of electrons. The search was on for the truth and beauty that might be represented in quarks twice as heavy, say, as the charmed quark. If they were created in pairs, like the charmed quarks in the gipsy, they might need 6000 mass-energy units or more to produce them.

In 1975–6 a big experiment was in progress, using the full force of the giant accelerator at Fermilab, to look for new forms of matter, of up to 20,000 mass-energy units. The team from Columbia University, Fermilab itself and the State University of New York at Stony Brook, was under the leadership of Leon Lederman. The experiment was similar in principle to Sam Ting's in the discovery of the gipsy, except that the energy was greater.

By February 1976 they thought that they had found something: a cluster of events at 5970 mass-energy units. Hearts beat faster – could this be truth or beauty? But the discovery was not confirmed. The Fermilab team could comfort itself that it still had a great new ocean to explore, all the way up to 20,000 mass-energy units, where particles as heavy as oxygen atoms might exist.

Doubts also overtook Stanford's heavy lepton, the U particle. There were suggestions, notably from the DORIS experiments in Hamburg, that it was an illusion – perhaps a misinterpretation of the decay of the charmed particles. At the time of writing these doubts have not been resolved and they illustrate again the difficulties and tensions of high-energy physics. A fair summary of the situation may be that there is no very compelling evidence so far for nature deploying more than four types of quarks and four members of the electron family.

The ultimate decay

In this account of the quarks so far I have tended to follow the 'standard' theory, which means in effect the theories favoured by Murray Gell-Mann, Sheldon Glashow and their young confederates. But it is probably fair to say that the 'heretics' at least matched the 'orthodox' in numbers and the threatened population explosion among the quarks was one of the reasons for unrest.

With or without truth and beauty, there were already too many basic particles of matter for any theorist to feel wholly contented. The electron family was not so bad. If nature chose to play games with the masses of heavy and superheavy electrons, that was a long-standing cause of puzzlement but the family remained reasonably compact.

The quarks were untidier because the theory of the colour force meant they had to be multiplied by three, since each type of quark could have any of the three colours. Counting only as far as charm, that meant twelve altogether: twelve particles each said to be a fundamental constituent of matter.

Perhaps the quarks were *not* basic particles, but combinations of other particles. Various theorists explored this possibility, including Moo-Yong Han (Duke University) and Yoichiro Nambu (University of Chicago). In an Indo-Pakistani alliance, Jogesh Pati (University of Maryland)

and Abdus Salam (Trieste and London) developed the idea strongly from 1973 onwards.

The Pati-Salam scheme focused on the qualities of matter represented in the quarks. It promoted the principle of 'one charge, one particle', where 'charge' in this context means qualities like strangeness and redness. The idea was that each quark consisted of a combination of two 'pre-quarks' or 'preons', one defining the type of quark (up, down, strange, charmed) and the other its colour (red, green or blue).

The electrons were brought into a unified scheme by saying that they possessed a fourth colour (lilac). Individual members of the electron family matched up with particular types of quarks, so that the heavy electron, for instance, became, in the Pati-Salam picture, a strange lilac quark. Then all of the particles could be made with just eight pre-quarks, four for defining the different types of quark and four for colour. That halved the number of 'basic' particles.

This heretical theory of Pati and Salam had interesting implications. One technical point of interest was that, if the quarks were made of pre-quarks, they did not need the 'fractional' electric charges assigned to them in the basic Gell-Mann scheme. They could have the same charges as the electron and anti-electron.

More significantly, if nature worked that way, the taboo against naked colour and the appearance of single quarks could be broken. In fact, when Martin Perl at SPEAR reported the simultaneous production of electrons and heavy electrons, and interpreted it as signifying superheavy electrons, Salam had a counter-suggestion. He argued that the effect could just as well be produced by red quarks appearing in public for the first time.

Bringing quarks and electrons together into one family implied that quarks could break down into electrons. That aspect of the Pati-Salam scheme was more readily acceptable than others, to their fellow-theorists. Indeed any theory that linked the quarks and electrons in a satisfactory manner might have the same remarkable conclusion: that the protons, the nuclei of hydrogen and the raw material of the universe, were not stable.

That crazy idea became respectable more quickly than most, even though experimental confirmation was not readily available. It offered a satisfying answer to a cosmic riddle. 'Why, out of all the heavy particles in the universe, does the proton alone exist for ever?' The reply was: 'It doesn't!'

Sooner or later, the proton would break down into an anti-electron and several neutrinos. The lifetime of a single quark in the Pati-Salam theory was a billionth of a second or less. But the three quarks making up a proton lent one another mutual support: only if all conspired to decay at the same moment would the proton actually break down. That had to be an extremely rare event, otherwise we should be highly radioactive on account of protons decaying in our bodies.

Salam estimated the life of the proton at about a thousand billion billion billion (10^{30}) years, a figure that gave no cause for alarm! As it happened, experiments had attempted to measure the lifetime of the proton, by detecting extremely rare breakdowns in a mass of material. They had already put lower limits on the proton's life-span, at around that very value. Salam was therefore hopeful that experiments would, before long, give an actual measurement of the decay of the proton.

The ultimate task of the weak force as Cosmic Alchemist was, in this theory, the abolition of all heavy matter. The growing conviction that the proton was unstable marked the final retreat from the old belief in the incorruptibility of atoms. Around 1900 the students of radioactivity realised that atoms could change into different atoms. A few decades later, uncertainty and anti-matter were working their looms of creation and annihilation, but the normal proton in the normal stable atom seemed secure in perpetuity. By the 1970s even that surrogate for the perfect atom was compromised. The only stuff in the universe that looked reasonably secure in its existence was the swarm of neutrinos.

Pati and Salam envisaged at least one more 'layer' of matter beneath the quarks. Others thought that, in principle, the subdivision of matter must go on for ever, but that idea stemmed from philosophy rather than physics. Chairman Mao, for instance, was said to have taken that view and it was favoured among Chinese high-energy physicists, who called quarks 'stratons' to indicate that they were just a layer or stratum in the scheme of the universe. But in the West there was a widespread suspicion that quarks were the bedrock.

The end of the road?

To some physicists the idea of quarks and colours that could never be observed directly was altogether too reminiscent of Lewis Carroll's White Knight:

> 'But I was thinking of a plan
> To dye one's whiskers green,
> And always use so large a fan
> That they could not be seen.'

The proper retort available to the literate quark-slaver was noted by Victor Weisskopf of the Massachusetts Institute of Technology. It was in the King James Bible, where an unusual translation of *Hebrews* 11: 3 produced a cosmological observation by St Paul: 'Things which are seen were not made of things which do appear.'

The apparent refusal of the quarks to show themselves outside the proton was to some leading theorists the most significant fact about them. It could be adopted as a point of principle and be understood in terms of the colour force and long-distance slavery. But philosophically it seemed important, too. As Sheldon Glashow said: 'Maybe nature is giving us a signal that we've come to the end of the road.'

Human beings had explored deeper and deeper into the micro-universe, finding the atom in the molecules, the nucleus in the atom, the proton in the nucleus, the quark in the proton. At each previous stage, you could dig out the new object and have a look at it. If you could not do that with a quark, it might be futile to ask what the quark itself was made of: indeed, traditional ideas about the meaning of existence seemed to break down at this stage.

If that reasoning was correct, you could only look to the patterns of the quarks themselves, in trying to understand why nature chose to fill the universe with quarks – and why it provided all those types and all those colours. So, assuming that quarks really could not be liberated, and that no evidence would be forthcoming of other particles inside the quarks, what was the way ahead?

According to Murray Gell-Mann there were two main possibilities. One was to find some elegant mathematical scheme in nature that required just the large numbers of particles – quarks of different types and colours, and the members of the electron family – that were discovered. If an overall unified theory were to succeed, it had to have that character. There were certain complicated 'groups' known to the mathematicians which might fit the bill. As for the other main possibility, Gell-Mann asked: 'Will some unknown young scientist find a new way of looking at fundamental physics that clarifies the picture and makes today's questions obsolete?'

Meanwhile everyone had reason enough to be gratified with the way theory and experiments in high-energy physics had moved forward together in the 1970s. Before that, the theories of particles and forces were fraught with perplexities and there was no clear direction in which to go. Experiments had only the general search for quarks as a major objective within the power of the available machines. But from 1971 onwards everything went right. Within five years the perception of the micro-universe was enriched to an astonishing extent. Above all, the behaviour of particles and forces was making good sense. Charm was perhaps the best token of that, because the need for it was fairly plain even before the gipsy was discovered.

Yet a great flaw remained. The electric, weak and strong nuclear forces were well encompassed by high-energy particle physics. Another cosmic force, gravity, was missing from the scheme. Whether or not the quarks were really the end of the road in the subdivisions of matter the theories were simply incomplete without gravity.

It was the most familiar force: gravity detectors in the head enabled human beings to walk upright, and the thickness of their bones was adjusted to oppose the gravity prevailing 6000 kilometres from the centre of the Earth. And gravity was the most truly cosmic of the forces. It ruled the wide universe, despite its inherent weakness, because of its unlimited range and its capacity to act on all matter and energy equally, without regard to electric charge, colour charge or any other quality. It was also 'single-minded'. People often dreamed of devices in which one object might repel another, by anti-gravity, but gravity remained stubbornly attractive in all circumstances.

Connecting gravity with the other forces was far from easy. In high-energy experiments, the gravity between particles was far too feeble to notice, never mind measure. The theoretical difficulties were well illustrated by Albert Einstein's failure. He spent the last thirty years of his life attempting to unify gravity and the electric force, and made no significant progress. But when Einstein's own ideas about gravity were taken up and re-examined by others who did not share his mistrust of the ideas of the particle physicists, hints of a possible connection began to appear in an unlikely context.

5 Starcrusher

Black holes gripped the thoughts of the rising generation of theorists as arguably the most remarkable objects of which the human mind had yet conceived. They were bottomless pits of gravity, said to be dotted about the universe, such that anything falling into them would lose all contact with the rest of space and time. In them, gravity might crush stars to nothing, consigning them to oblivion for ever.

No, not quite for ever. That was an important revision of the ideas about black holes, made in 1974. Eventually a black hole could give rise to a monstrous explosion, annihilating itself and spewing matter and energy back into the universe. And in that statement was wrapped one of the most far-reaching unifications ever attempted, of different phenomena in the universe.

Theorists in the leading centres concerned with Einsteinian issues abandoned their own inquiries to work out the implications of the new idea. Some enthusiasts did not hesitate to compare it with Albert Einstein's $E = mc^2$, which unified matter and energy. Whether you looked at it enthusiastically or sceptically, it was undeniably a daring leap of the human mind into new realms of thought.

This chapter tells of the study of gravity, which was going on in parallel with the advances in understanding particles. Gravity could appear weak to the point of being negligible in experiments with small objects. Yet it was plainly the master of the universe and the maker of galaxies, stars and planets. And when matter was closely packed together gravity could overwhelm all the other forces and ride roughshod over many familiar cosmic laws.

Growing numbers of astronomers called themselves 'high-energy astrophysicists' because, in many objects in the universe, they could see matter squirming under extreme conditions of heat, magnetism or gravity. The parallels between high-energy research on man-made machines and among the stars were striking enough. Nature had its own accelerators in the form of varying magnetic fields that caused particles to orbit and gather speed. For the practitioners there was the same obligation to make bold inferences about the cosmic forces from a smudgy photograph or a run of instrumental measurements.

Exchanges of ideas and knowledge between the investigators of the

instrument for studying
losions in distant galaxies.
radio telescope at
sterbork in the Netherlands
sists of 12 identical dishes.
two nearest ones are on
s, so that the spacings can
varied; as the Earth turns,
telescope sweeps out a
hering region equivalent to
huge dish. What goes on
he heart of an exploding
axy may be a crucial test
the theory of gravity.

Albert Einstein (1879–1955) was born at Ulm in Germany and began working his revolution in physics while an inconspicuous patent examiner at Bern in Switzerland. Strange to say he won his Nobel Prize for a discovery away from the main stream of his theoretical work: for realising that the way in which light knocked electrons out of metal meant that light consisted of particles. His theories of relativity, special and general, were his greatest gifts to physics. In the late 1920s with the rise of 'uncertainty' in the atomic domain, Einstein was appalled by the apparent allegation that 'God played dice'. A sharp intellectual power-struggle ensued between Einstein and Niels Bohr, which Bohr won decisively. Being a famous Jew, Einstein was singled out for attack by the Nazis. He left Germany in 1932 and settled in Princeton. In 1939 he signed a fateful letter to President Roosevelt drawing his attention to the possibility of a nuclear bomb and the German interest in it. Towards the end of his life, Einstein declined the invitation to be President of Israel.

broad universe and the micro-universe of atoms and particles had always been fruitful. The crucial element helium was discovered by its characteristic light coming from the Sun. The processes by which stars burned and made the chemical elements were worked out in a joint operation using telescopes and nuclear accelerators. Cosmic rays arriving at the Earth gave information both about particles and about the Milky Way. But there remained a clear division of labour as far as the cosmic forces were concerned. While the electric, strong nuclear and weak forces were suitable for investigation on the ground, the best opportunities for studying gravity lay in the massive bodies of the universe.

The twentieth-century theory of gravity was devised by Albert Einstein in 1916, but not until the 1970s was it put rigorously to the test, or its full implications worked out. For half a century gravity seemed to exist in a compartment quite separate from the other cosmic forces, and that remained the most obvious impediment to a grand theory of the universe. The reasoning about exploding black holes showed how gravity might manufacture particles. It made hopes rise.

Gravity after Einstein

The first men to stand on the Moon, Neil Armstrong and Edwin Aldrin, had the task of positioning a reflector there. It enabled American scientists to shoot a laser beam from the Earth to the Moon and detect the reflected light back on Earth. By that means they could measure variations in the distance of the Moon to an accuracy of 30 centimetres. After six years and 1500 measurements they were able to announce that Albert Einstein's theory of gravity looked good.

One of the signatories of the report that said so was, ironically, Robert Dicke of Princeton. He and Carl Brans had challenged Einstein's theory and offered an alternative of their own. But again and again the predictions of Einstein's theory of gravity were being borne out by increasingly stringent tests while its rivals' predictions were failing.

The search for the key to the universe had thrown up dozens of theories of gravity, but only two ever seemed successful. The first was Isaac Newton's, which reigned for more than two hundred years. The second usurped the throne quite suddenly. Einstein worked it out by 1916, in the midst of the Great War. By 1919 his predictions of the effect of gravity on starlight had been confirmed during an eclipse of the Sun.

Einstein's theories of relativity were a model of excellence for anyone who wanted to find logic in the workings of the universe. Basic considerations of the need for self-consistency in nature, whether observed by

136

earthlings or the inhabitants of another galaxy, led him to his highly original ideas. One of them was that matter deformed space-time, so that particles of light did not travel in ideal straight lines between objects. The effects of the curvature of space were equivalent to the force of gravity.

This theory of gravity, known ceremoniously as 'general relativity', was a *tour de force*. He found a new way of thinking about gravity, and about space and time. He brought new kinds of mathematics into physics. And he presented his colleagues with a package, neat, complete and all ready to be tested for its veracity. Compare that with the way hundreds of other able men were later to agonise over the weak and strong forces, inching their way forward decade by decade, and Einstein's solo effort shone very bright.

But he flew solo only while generating the crazy idea that space-time was curved. He was a physicist, not a deductive philosopher or even a mathematician. He knew all the strengths and the very few weaknesses of Newton's theory; and he benefited from results from many experiments. Not the least of these was a conclusion, from two centuries of work, that the 'mass' which made an object resist acceleration was very closely, probably exactly, the same as the 'mass' on which gravity acted.

To cast Einstein's theory in simple yet modern terms, it was a 'gauge theory', like those of the other cosmic forces we have encountered. The force-carrying particles, the gravitons, felt the force of gravity themselves. Powerful ideas of symmetry ran through the theory. Salient was the notion that the laws of nature should appear the same to any two observers in the universe, regardless of how they were moving or accelerating relative to one another.

The gravitons felt the force of gravity themselves simply because they possessed energy – which Einstein's earlier work had shown to be equivalent to a mass. As a result of their interactions, the gravitons did not travel in classical straight lines. Nor did light. In effect, the space around a massive object acted like a distorting lens.

In speaking of curved space, Einstein was not adding any mysterious new dimension to the universe. He simply took the three dimensions of space and added time, as measured by the distance light could travel in a given interval. He did, it is true, churn them together in an unconventional way. There were curious effects on time: for example, a clock taken to the visible surface of the Sun would run slow by about a minute a year, because of the Sun's strong gravity.

As no one could conveniently think of three-dimensional space curved into a fourth dimension, the rubber-sheet analogy became popular. You imagined space as being a flat sheet, with the stars as weights which

Curved space. To the three dimensions of ordinary space Einstein added the fourth dimension of time, as the distance travelled by light in a given time. That fourth dimension made possible the distortions of space by the mass of stars and other objects. It is hard to visualise unless you flatten space in your mind and then imagine the stars curving it into another dimension, as shown here. Around the star a planet orbits like a racing car on a banked track.

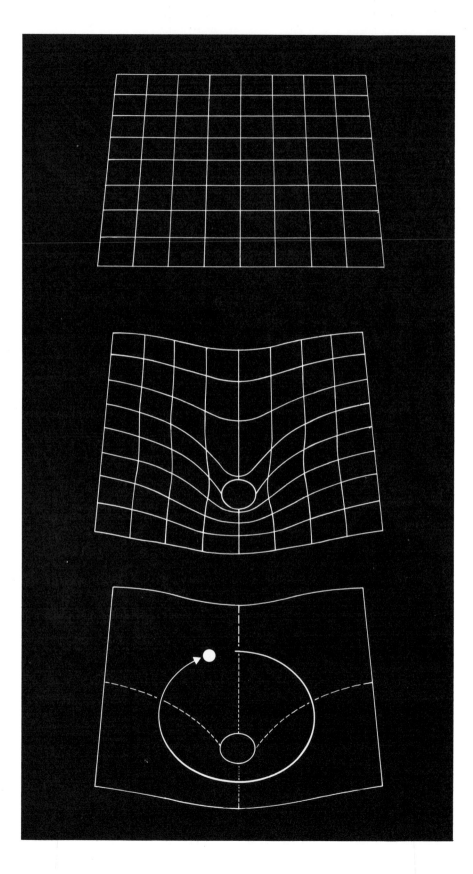

distorted it. The planets held in orbit around the Sun were then like racing cars turning on a banked track. Because of the distortion of space, the Sun's visible surface was a few million square miles smaller than you would expect from its diameter.

Einstein was sufficiently pleased with his theory to remark that he would feel sorry for God if it proved wrong. But he was careful to offer to his colleagues some means of checking it. One was the deflection of starlight by the Sun, already mentioned. Another was the matter of clocks running slow near a massive body, which should have the effect of reddening the light coming from stars, because the atoms which produced the light would also be running slow. That proved difficult to check. Eventually ingenious experiments in Britain and the USA showed the same effect on Earth, with nuclear 'clocks' running slower if they were closer to the ground. But for half a century Einstein's theory stood like an admired but neglected monument, aside from the main interests of the physicists. Very rarely did the differences between Newton's and Einstein's versions of gravity seem important, because gravity was seldom strong enough for them to be of any great consequence.

A rapid succession of astronomical discoveries in the 1960s helped to force the revival of interest in Einstein's theory. They all pointed to the existence of very dense agglomerations of matter. First came the quasars and the realisation that compact, starlike objects, pouring out radio waves and intensely blue light, lay at very great distances. That meant they had to generate enormous energy from a tightly packed mass, where gravity would be very strong. Gravity itself was a prime candidate to be the supplier of energy.

The discovery that empty space was warm followed quickly. It was the best evidence that the universe began in a Big Bang, which in turn implied that all the matter and energy of the universe was concentrated together in the first few moments. You needed the best possible theory of gravity to try to comprehend the conditions prevailing in the Big Bang. Finally, the identification of pulsating radio sources ('pulsars') as neutron stars brought astronomers face-to-face with matter in a state half-way to a black hole. In a neutron star, gravity had already overwhelmed the electric force and crushed the atoms completely.

Checking up on Einstein then became a matter of urgency. Already a few pioneers were trying to detect gravitons, the gravitational force-carriers. Gravitons would be extremely feeble things, and there was no hope of detecting any from the commonplace operations of gravity. It was expected, however, that strong bursts of gravitons, for example from stars collapsing under gravity, might set sensitive detectors vibrating on

139

Earth. The first claims of success came from the University of Maryland in 1969, but they were not confirmed. Various laboratories around the world busied themselves with the design of even more sensitive instruments.

Nature offered another opportunity of a different kind, with an extraordinary state of affairs announced in 1975 by astronomers using the radio telescope at Arecibo in Puerto Rico. It was a pulsar, a neutron star emitting extremely regular radio pulses (an excellent natural clock). And it was circling extremely closely, every eight hours, around another object, an unseen collapsed star, and therefore subject to very strong gravity. Observing the pulsar for several years might yield a new test of Einstein's predictions.

But these possibilities, even the last, were relatively mild tests of gravity. Going to extremes was often the best way to check a theory, and to make sure you really understood how a piece of the universe worked. An extreme state of gravity was always implied by Einstein's theory. As Karl Schwarzchild noticed as early as 1917, lurking in the equations was a black hole.

'Black holes are out of sight'

If the poets were right in saying hell was black the physicists glimpsed it in their theories. At first black holes were just another crazy idea – a figment of the human imagination. When later they seemed to be real members of the universe's population of starlike objects, they troubled the imagination even more.

The original idea of a black hole was simple enough. In the eighteenth century the Marquis de Laplace reasoned that, if light could feel the force of gravity, a massive enough star would choke off its own light, preventing its escape into space. The star would then be blacked-out from the view of any outside observer: a curious enough idea, well ahead of its time. Two centuries later American astronomers had stickers on their cars that read: 'Black holes are out of sight'.

When gravity drew materials together to make a star or a planet, other forces came into play which resisted gravity. The Earth obtained its precise size from the balance struck between gravity and the electric force in the atoms of the rocks and in the Earth's core of molten iron. In an ordinary star, like the Sun, gravity was much stronger but the gassy material resisted compression because it was very hot, and continued that way because the core was burning hydrogen (in the nuclear fashion) and producing energy.

ck hole in curved space.
e the distortion of space-
becomes very great. But
flattened-space picture
uld not be taken too
usly. In practice, a black
would be a small,
erical object, which one
ht pass around in any
ction.

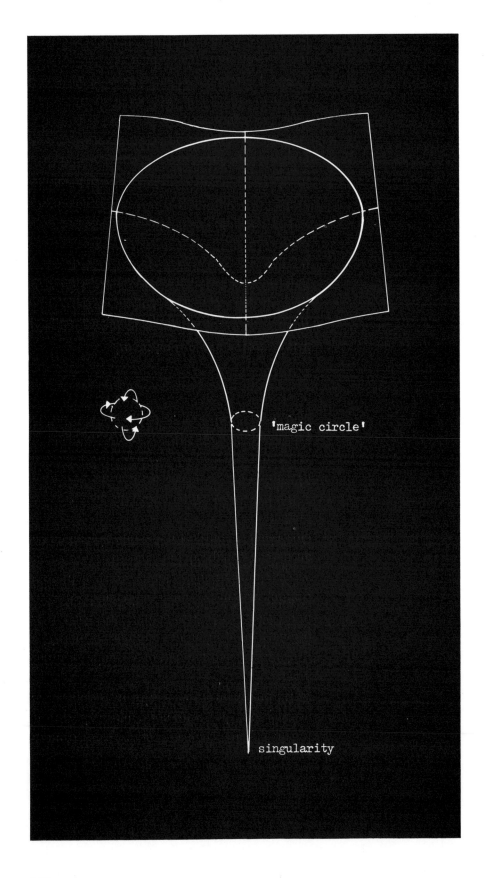

'magic circle'

singularity

The astronomers knew that, when an ordinary star ran out of energy, a series of dramatic enlargements and puffs occurred, but the outcome was that the once-gassy star shrank into a solid body, far smaller than before. Some astronomers said, fancifully, that the star became a diamond in the sky – diamonds being crystals of carbon and carbon an important product of a senile star. The star was a 'dwarf', roughly the size of the Earth, but much denser.

The demise of a star more massive than the Sun could compress the central remains of the star violently enough to defeat the electric force. The remaining line of resistance to gravity was then the strong nuclear force. Theorists envisaged a star like one enormous atomic nucleus, and called it a neutron star. Electrons were in a sense squeezed into the protons to make neutrons, the uncharged relatives of the protons. The pulsars that turned up in 1968, flashing rhythmically many times a second, were very small stars associated with the remnants of the explosions of massive stars. They packed a mass greater than the Sun's into a ball only 20 or 30 kilometres in diameter. If you thought of standing on a neutron star you would weigh more than a million million tons and would be at once flattened into a pancake of atoms.

How did nuclear matter behave under pressure, in a neutron star? A theoretical picture emerged of a solid crust subject to 'starquakes' which altered the rate of spinning of the neutron star. (Such 'glitches' were detected by the radio astronomers.) The crust enclosed a dense liquid of neutrons, under extremely high pressures. The core of a heavy neutron star might be a solid mass of neutrons, or perhaps a completely new state of matter. The pressures that nuclear matter might withstand depended on precisely what form it took, but it seemed unlikely that a neutron star would survive if it were more than about twice as massive as the Sun.

Even before the discovery of the pulsars, theorists of gravity had taken the argument a step further. If a star more massive than those which made the neutron stars were to collapse, gravity could overwhelm the nuclear forces too. Then the star would not stop shrinking at the size of the neutron star but would go on collapsing. Its gravity would grow more and more intense, until it could prevent all light escaping and made a black hole. The star effectively disappeared from the universe, except for the intense gravity that surrounded the place where it disappeared. The rubber sheet of space-time was distended in a narrow, very deep well. But the analogy was misleading, because you could circle as freely around a black hole as around any other object in space.

You might have reason for doing so, if you wanted to travel through time. The strong gravity churning space and time near a black hole would slow down clocks and life processes so greatly that the lapse of ten thousand years on Earth could seem like only a few weeks to the voyager orbiting the black hole. But there would be no going back in time, to return to your natural time-frame. And if you were careless enough to fall into the black hole, you could not return to Earth through space, either.

There would be a 'magic circle', actually a sphere around the vanished star, but a dire perimeter not marked by any boundary posts. Anything falling through it would never come out again, because nothing could travel faster than light. For a black hole twice as massive as the Sun, the 'magic circle' would be about 12 kilometres in diameter.

Here indeed was an 'undiscovered country from whose bourn no traveller returns'. And the fate of the imagined space-traveller who stumbled upon a black hole became a commonplace way of describing the extraordinary work of gravity, in and around a black hole. Before being trapped and crushed, the unwary astronaut would first be stretched into spaghetti.

The first hint of trouble might be his hair standing on end, his feet and hands feeling heavy, his head light. The astronaut's blood would drain into his limbs, bringing merciful unconsciousness before gravity rendered his body into meat, into molecules, into atoms, and eventually into a long beam of particles hurtling into the black hole.

Spaghettification was due to gravity intensifying, metre by metre, in the approach to a black hole. It was a tidal effect, an extreme version of the process by which the Moon would pull more strongly on sea-water immediately beneath it than on the oceans to the far side of the Earth. In the more severe conditions around a black hole, the force of gravity increased so rapidly towards the centre that it could easily be a million times stronger at the spaceman's feet than at his head. So he would be torn apart by the tide.

Gruesome calculations of that kind followed directly from Einstein's theory of gravity. The modern version of the basic theory of black holes emerged in the late 1960s, chiefly from the work of Roger Penrose (then at Birkbeck College, London) and Stephen Hawking of Cambridge University. If the black hole ran counter to familiar notions about the universe, a further consequence of the theory was a positive affront to them. It seemed that, in a typical black hole, nothing could halt the collapse of its contents. All the matter of a star forming a black hole (or any astronaut blundering into it) would be crushed by gravity into a geometric point, smaller than a pinprick.

A black hole could thus contain at its centre an even more remarkable object: the point-like 'singularity', where existence and even time itself seemed to cease. It offended many people's preconceptions of good behaviour on the part of nature. Even ten years later, familiarity with the idea had not subdued those who wanted alternative theories of gravity to prevent such total catastrophe, or who simply denied that singularities ever actually happened.

Penrose and Hawking were firm. If Einstein's theory of gravity was correct, first the black holes and then the singularities were unavoidable, whenever enough material was gathered together in one place. Any special forces that people tried to envisage to arrest the collapse, or cause the infalling material to bounce out of the gravitational pit, merely hastened the doom. The reason was not mysterious. The exertion of any counter-force implied the existence of force-carrying particles of a certain mass-energy, which simply added to the total force of gravity.

Plenty of big stars seemed massive enough to make black holes, even if most of their substance were scattered into space when they exploded. So the universe ought to be as full of holes as a colander. If astronomers could find them and study them closely, they might provide an exceptional way of checking Einstein's theory. By definition, you could not expect to see a black hole directly. But you might detect its presence by the unusual behaviour of objects falling into it. That became one of the first objectives of the fledgling science of X-ray astronomy.

A gurgle of X-rays

The visible universe looked misleadingly serene because human eyes were tuned to the emissions of the Sun, a well-behaved, long-lived star. From scenes of cosmic upheaval and violence, the strongest emissions were invisible rays. And the maturing of X-ray astronomy in the 1970s resulted in a crop of surprises not unlike those produced by the advent of radio astronomy a generation earlier.

The astronomers looking for X-rays coming from the universe had to lift their detectors above the Earth's atmosphere. The X-rays that pass so easily through human tissue are, curiously enough, blacked out by the air. The first X-ray star, Scorpius X-1, was detected in 1962 during the short flight of an unmanned American rocket, equipped with a detector by Riccardo Giacconi, of Cambridge, Massachusetts.

X-ray astronomy matured with the arrival in orbit of the satellite *Uhuru*, the first durable observatory purpose-built for studying the universe by X-rays. Conceived by Giacconi, this American satellite was launched from a platform standing in the sea off the coast of Kenya, in December 1970; it was the anniversary of Kenyan independence and

Uhuru is the Swahili word for 'freedom'. It mapped about 170 X-ray sources in the sky. Its detailed watch on individual objects gave the first reliable clues to what X-ray stars might be.

In the years that followed, further X-ray satellites kept up the pace of discovery: *Copernicus* in 1972, an American astronomical satellite equipped with X-ray telescopes developed by University College London; *ANS*, the Astronomical Netherlands Satellite, and *Ariel 5*, a British-born satellite, both launched by American rockets in 1974; *SAS–3*, successor to *Uhuru*, launched in 1975. A new generation of more sensitive instruments was designed for the satellites of the late 1970s.

X-ray stars had a tendency to flare up or vanish, so that a dozen of *Uhuru's* stars were missing when *Ariel 5* looked for them a few years later, and a similar number of new ones had appeared. *Ariel 5* was crammed with instruments. Three were devised by University College London, where Robert Boyd and his colleagues had developed new techniques for X-ray astronomy and tried them out in *Copernicus*. Two other instruments came from Leicester University and one from the Goddard Space Flight Center in the USA. On command, puffs of gas swivelled the satellite to point several of the instruments at a chosen star or galaxy, while survey instruments on the side scanned right around the sky as *Ariel 5* spun about its axis.

It was strange to be able to walk into a room at a British university (at Leicester, say) and gather the latest news of extraordinary events going on far away in space. The X-ray astronomers' hot-line to the universe linked them to their satellite via ground stations across the world. And by that route Kenneth Pounds and his colleagues first 'saw' an X-ray source brewing up amazingly. *Uhuru* had recorded a very faint X-ray source in the constellation of Monoceros. In August 1975 the Leicester survey instrument in *Ariel 5* detected it flaring. The X-ray star multiplied its intensity 3000 times in ten days, to become by far the brightest X-ray object in the sky. Then it faded away, with just a flurry or two, during the following weeks and months.

Ariel 5 did not get a very precise position for it. But the team at the Massachusetts Institute of Technology quickly turned *SAS–3* to look at it. With the better fix, a consortium of astronomers using an optical telescope on Kitt Peak, Arizona, was able to find the flaring star, by visible light, in just a couple of nights. As the source of the explosion of X-rays in Monoceros, the observers identified a star, fairly faint but very much brighter than it appeared in an earlier photograph of the same patch of sky.

A 50-inch telescope recently transferred to the new McGraw-Hill Observatory on Kitt Peak, made the identification. It carried sensitive electronic detectors for registering faint sources of light. Its chief advantage, though, was that of being the only sizeable telescope available full time for studying the visible counterparts of X-ray sources in the universe, and able to keep up with the pace of events recorded by the X-ray satellites.

Astronomers had seen that star in Monoceros flare up before, in 1917. Now it was active again, ten thousand times brighter than normal. Other observers judged it to be about 6000 light-years away. If so, energy was pouring out of the eruption at so great a rate that, in the judgement of some astronomers, the event in Monoceros demanded the presence of a companion black hole, perhaps six times as massive as the Sun. The flare-up would then be due to material from the stars falling into the black hole.

By that time there was already a strong consensus of opinion that another X-ray star, Cygnus X–1, probably contained a black hole. A fairly clear picture of 'ordinary' X-ray stars had emerged, from the *Uhuru* findings. They were remarkable enough objects, representing a phase in a cycle of events whereby the material of one star could be decanted into another star.

X-ray star: matter falling from one star to another

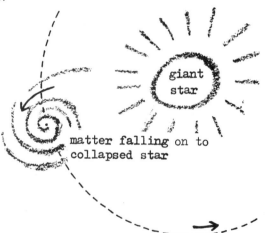

Imagine a pair of stars orbiting closely around each other, one of them being very small. The X-ray source Centaurus X–3 was found to pulse like a pulsar every 4.8 seconds, while it orbited every two days around a great star fifteen times more massive than the Sun. The small star, evidently a neutron star, acted like a vampire, sucking or at least capturing gas from the other star and causing it to crash on its own surface. Under the enormous force of gravity, the falling material became hot enough to emit X-rays, while the neutron star funnelled it on to its magnetic poles. The pulsations seen from Earth were then due to the spinning of the neutron star, which pointed its hot magnetic poles alternately towards the Earth and away from it.

iel 5, an X-ray satellite
nched in 1974. Several X-
instruments look out along
axis of the spacecraft;
nspicuous on the side is a
nel of four detectors which
n the sky as the spacecraft
ns. This instrument
eicester University Sky
rvey) detected an
raordinary outburst of
rays from a star in the
onoceros constellation
the summer of 1975.

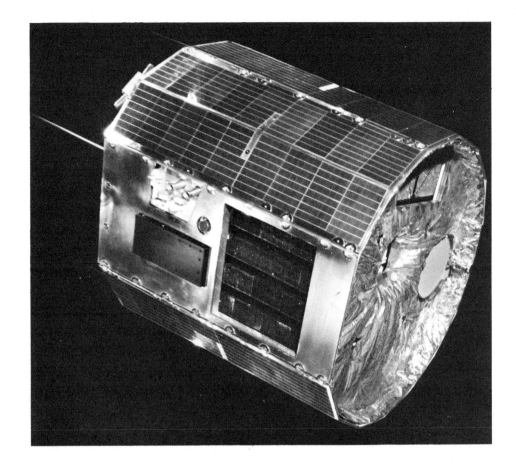

oming in on a suspected
ick hole. The rings and
xes correspond to the
certainties in the position of
X-ray star Cygnus X–1, in
ation to visible stars. 'ASE'
d 'MIT' refer to positions
und by X-ray telescopes in
ace (American Science and
gineering Inc., and
assachusetts Institute of
chnology). In 1971 a
taclysm in Cygnus X–1
used it to start emitting radio
ves and as a result radio
tronomers at the US
ational Radio Astronomy
servatory (NRAO) and
esterbork (W) in the
etherlands were able to
npoint the bright star HDE
6868. The supposition is that
black hole is in close orbit
ound it.

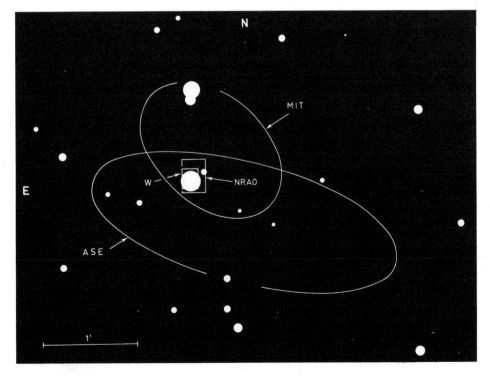

**X-ray pulsar
has hot spots
and flashes
like a lighthouse**

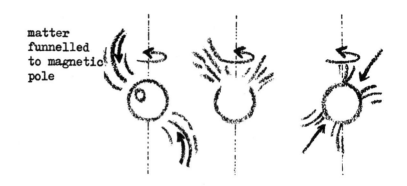

matter
funnelled
to magnetic
pole

Cygnus X–1 did not fit that picture exactly. At its site, astronomers could see a big blue star about 6000 light-years from the Earth and again with about fifteen times the mass of the Sun. The X-ray record showed that the source of X-rays was orbiting around it every five and a half days. It seemed to be heavier than you would expect for a survivable neutron star. Moreover, there was no regular pulse in the X-ray emissions, but they fluctuated rapidly and erratically.

The favourite theoretical model that emerged for Cygnus X–1 was of a black hole into which material from the big star was swirling, like water down a plughole. The infalling material formed a disc around the hole and spiralled inwards, becoming hotter and hotter, until it disappeared into the black hole, emitting a gurgle of X-rays. The fluctuations were then due to individual 'lumps' of material falling to their doom.

**Matter spiralling
into black hole
becomes 'X-ray-hot'
before it disappears**

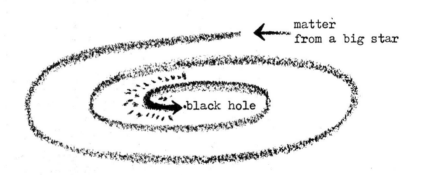

matter
from a big star

black hole

Theorists who disliked the idea of a black hole were able to produce other schemes to account for the behaviour of Cygnus X–1. Yet the black-hole theory was the simplest, once you accepted the premise. There was anxiety to find other possible 'black-hole candidates' among the X-ray stars to confirm the reality of black holes and perhaps later to test Einstein's theory of gravity in extreme conditions, by interpreting the details of how gas fell into the black holes.

The material of a star could finish up inside a black hole at least three distinct ways. It could collapse at the end of its life, into a black hole of its own making. Or a vampire-like black hole in orbit around it could drain its substance away. Thirdly, it might encounter a black hole and fall into it. In the wide open spaces of a galaxy, the environment of the Sun for instance, the chances of that happening were no greater than the chances of any other kind of collision between stars. The detection of X-rays coming from 'globular clusters' of stars suggested that such disasters might indeed occur in those densely-packed regions. And the wholesale swallowing of stars into a black hole could be fairly commonplace at the centre of a galaxy.

Gyros for exploding galaxies

Some objects in the universe were so extremely powerful that nothing except very heavy black holes seemed likely to account for them. X-ray astronomers detected powerful X-ray sources lying across the ocean of space and time, where great galaxies, each of them comparable with our own Milky Way and containing many thousands of millions of stars, were racked by explosions. But exploding galaxies had been known and investigated for a long time, by the radio energy they emitted.

The first 'radio galaxy' was spotted in 1946 and thereafter a large part of the story of radio astronomy was the gradual elucidation of the nature of such objects. Some sources of natural radio emissions turned out to be small, leading to the discovery of the quasars. Amid a lot of controversy, the conviction grew that quasars and the much broader radio galaxies were closely related to each other, and also to other galaxies in various peculiar hot, bright or distorted conditions. It seemed that the core of almost any galaxy, perhaps including our own Milky Way, was capable of staging explosions far surpassing in intensity and duration the explosion of a massive star.

All kinds of theories were advanced to account for these eruptions. Nuclear explosions were quite inadequate. If one was not to invent a new cosmic force, only gravity seemed capable of generating the necessary energy. In the 1970s the simplest and yet most provocative explanation was that the galaxies concealed very massive black holes. Vast quantities of material, whole stars probably, could fall into such a gravitational pit, creating an enormous central explosion as they went.

By that time, the British (at Cambridge) and the Dutch (at Westerbork) were using radio telescopes specially adapted to the task of exploring the details of radio galaxies. Each telescope consisted of a long row of radio-receiving dishes which stayed locked on a particular object in the sky, while the Earth turned. By elaborate comparisons of the arrival-times of radio energy at the different dishes you could map an object as precisely as if you had one enormous dish.

The resulting maps of radio galaxies showed some well-known features, notably the great lobes of radio emission that lay on either side of the visible galaxy. But high-precision mapping also showed 'hot spots' in these lobes and in the central galaxy itself. Sometimes it was plain that a single galaxy had exploded on several separate occasions.

Why did radio galaxies have the curious 'two-eyed' or double appearance? Roughly equal lobes of radio emission extended in opposite directions from the visible galaxy, which plainly generated them. As millions of years passed, the lobes shifted further and further apart.

Astronomers at first supposed that the lobes were twin clouds of hot gas expelled from the galaxy in one comparatively brief event, which then gradually gave off their energy as they moved away from the galaxy. But the more detailed maps encouraged the view that the galaxy was continually supplying the lobes with energy. One promising explanation came in 1974 from Roger Blandford and Martin Rees of Cambridge, who called it the 'twin-exhaust' theory. A spinning mass of gas containing a strong source of energy could, they said, spontaneously eject steady, narrow jets of fast-moving particles.

The jets would emerge in exactly opposite directions, along the axis around which the cloud of gas was spinning. These jets, travelling outwards through the thin gas in the space surrounding the galaxy, would then create a shock wave that generated the radio noise detected on Earth.

Behind the stars in the constellation of Leo Minor, and unseen except

Cygnus A, a 'two-eyed' radio galaxy. This powerful object (not to be confused with Cygnus X–1, the X-ray source) was the first radio galaxy to be discovered. This detailed map, with contours showing increasing radio intensity, was made in 1974 by astronomers using the five-kilometre radio telescope at Cambridge. The object in the middle corresponds to a visible galaxy, which has evidently expelled vast amounts of energy in opposite directions to make the radio 'hot spots'.

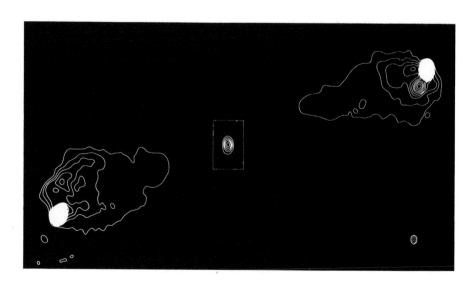

by radio, lurked the biggest and most powerful object of which human beings had any knowledge. It was so far away that the radio energy that astronomers were registering left it when microbes still ruled the Earth 1800 million years ago. Even at that vast distance its great lobes looked wider than the Sun, and its output of energy was immense. In the middle, and less than a hundredth of the size, lay a visible galaxy, the scene of some kind of catastrophe long ago.

Astronomers from Leiden used the Westerbork telescope to map this very large radio galaxy. They were fascinated to compare it with another map showing radio emissions from an object much closer and much smaller: the X-ray star Scorpius X–1. In both cases, two sources on either side of a central source lay in a neat straight line. It was tempting to think that nature could produce the same kind of fireworks on the scale of a galaxy as on the scale of a star. Would black holes account for it?

One very important clue came from galaxies that had exploded more than once. Successive pairs of gas clouds or exhaust jets came out repeatedly in exactly the same direction, even when the explosions were separated in time by millions of years. That fixity of direction was confirmed in what could be seen close to the very centres of exploding galaxies. But investigating the fine details called for something more than the Cambridge or Westerbork telescopes: you needed a telescope roughly as wide as the Earth.

The 100-metre radio dish near Bonn in West Germany, operated by the Max-Planck Institute for Radio Astronomy, was a formidable instrument in its own right – the world's biggest fully steerable radio telescope. But sometimes it acted as just one part of a much larger system, stretching from Germany to the Pacific Coast. Radio telescopes in West Virginia, Texas and California as well as Germany would all stare at the same spot in the sky for several hours a day and for several days.

Coordinated operations like that, using very long 'baselines', became fashionable among radio astronomers in different parts of the world,

'Twin-exhaust' for exploding galaxy

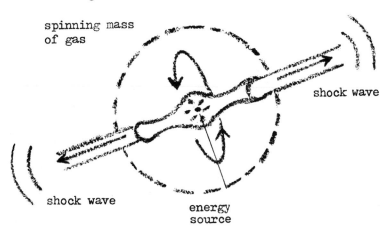

spinning mass of gas

shock wave

shock wave

energy source

The radio galaxy 3C 236, the largest object in the universe. It is 18 million light-years long. The 'radio photograph' below was made with the Westerbork telescope by astronomers of Leiden Observatory. The bright object in the middle corresponds with a visible, disturbed galaxy. It contains other, smaller radio sources aligned with the two conspicuous outer components, a fact which indicates that the explosive core of the galaxy 'remembers' a direction in space for many millions of years. One theory is that it incorporates a very massive, spinning black hole. The small central sources (map) were investigated with special radio-telescope combinations in the USA and in Britain. The 'blobs' are about 6000 light-years apart and the line running through them is the orientation of the much bigger, outer sources.

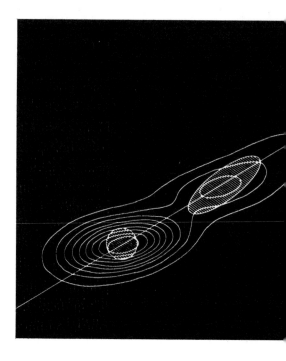

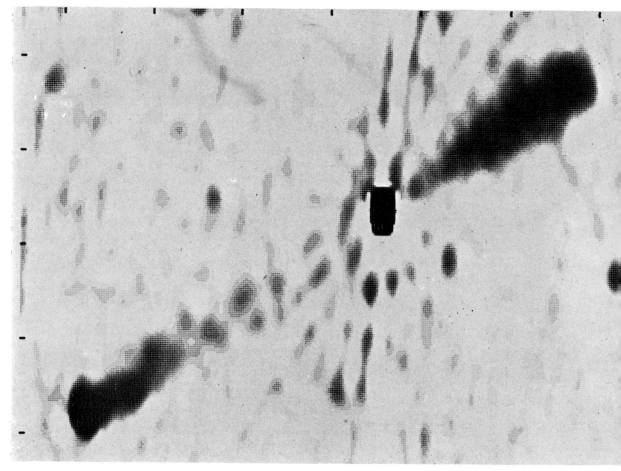

because the clarity with which you could observe an object in the sky depended on the overall width of your instrument. Their intercontinental baselines gave the radio astronomers far sharper 'vision' for small objects than even the grandest optical telescopes could provide. In 1974 the American and German collaborators peered into the very heart of the brightest and stormiest members of the Perseus cluster of galaxies – the galaxy known as NGC 1275. But the combined tape recordings from the four telescopes produced nothing like a picture. Instead the radio astronomers had to invent a form for the source of radio noise that would give exactly the same pattern of signals as those received by the different radio telescopes.

After months of work Ivan Pauliny-Toth at Bonn found the picture that served. In the middle of galaxy 1275, crowded close together, he put a row of three intense radio sources. The distance between the two at the ends was only about ten light-years. It represented an explosion that occurred in the mid-1950s, as events were recorded on Earth (in fact, galaxy 1275 was 400 million light-years away). Also deep inside the galaxy, though much wider than the triple source, was a double radio source found by American radio astronomers. It represented an earlier explosion. Around the galaxy a large lobe of radio emissions, mapped by the Westerbork telescope, corresponded with an explosion that occurred more than a million years before the most recent one. In all this, the extraordinary fact about such exploding galaxies reasserted itself. To the accuracy with which the astronomers could measure it (to about a degree of slope) the axis of the most recent explosion pointed in exactly the same direction in space as the two earlier explosions.

The only way anyone could think of, whereby the galaxy might pre-serve the memory of a direction so well and so long despite the most violent outbursts, was an extremely heavy gyroscope – a single rotating object, perhaps a million or a billion times more massive than the Sun. Pouring the mass of a million stars together might seem to be an infallible recipe for making a black hole.

Philip Morrison of the Massachusetts Institute of Technology argued that, if the material were revolving rapidly enough around the centre of the object, it could avoid collapsing into a black hole. A huge, spinning, flattened, highly magnetised star, that Morrison called a spinar, could then survive at the heart of the galaxy. It would not glow very strongly but could power explosions by flinging off particles and accelerating electrons by its intense magnetism. Indeed, Morrison offered spinars of various sizes to explain all kinds of events in the universe that otherwise suggested black holes.

153

But among other astronomers it became acceptable to speak of black holes a million or more times heavier than the Sun, inhabiting the centres of galaxies. Objects like that could gobble up gas and whole stars, and the energy released during the infall under gravity could quite easily create galactic explosions or a fierce glow. Stoking such a gravitational furnace with just one star a year would be sufficient to power a quasar. And the black graveyard of a million stars, surrounded by a swirling disc of disrupted stars, could also possess the gyroscopic spin needed to maintain a fixed axis in a long-lasting succession of explosions.

The evidence was inconclusive. A new generation of telescopes in space might confirm the reality of black holes, both in objects like Cygnus X–1 only a few times heavier than the Sun, and in galaxies like 1275 where the black holes would be far more massive. Nevertheless, the astronomers were already taking the extreme powers of the force of gravity very seriously, when they were invited to consider also the possibility of very small black holes, far less massive than a single star.

Gravity makes particles

Until 1974 the black holes were simply insatiable maws in space, capable of swallowing matter and removing it permanently from contact with the rest of the universe. Then one of the founders of black-hole theory came up with an exceptionally crazy idea – crazy in the admirable sense. At a meeting in Oxford, Stephen Hawking stunned his audience by declaring that black holes would explode.

A grave physical handicap made Hawking's work all the more remarkable. Stricken while an undergraduate at Cambridge, he fought a wasting disease of nerve and muscle that prevented him walking. In time, he could not use a pen either, so he had to do much of his complicated mathematics in his head. Speaking became difficult for him. But although the gentle gravity of the Earth confined him to a wheelchair, in his mind he wrestled with the overwhelming gravity of a black hole.

Hawking once remarked: 'If you have a physical handicap you just can't afford any psychological hang-ups.' Theorists are in any case paid to sit and think, so he had a very suitable trade to follow. And he found ample reason to be pleased with life, in his wife and children, in the medals and honours that his fellow-scientists showered upon him, and in the sheer excitement of ideas.

Even so Hawking's colleagues, who never doubted his talents, found increasing reason to be impressed by his courage. For a while close associates wondered if his illness was getting the better of him: he was

Stephen Hawking, leading theorist of black holes and how they could explode. His ideas provide an important link between theories of gravity and of sub-atomic particles.

Owens Valley

Effelsberg

igh-precision radio
tronomy, using
tercontinental combinations
telescopes. The
mbination shown below
chieved the analysis (above)
the three very small radio
urces at the core of the
laxy NGC 1275 (above,
ght), which evidently
ploded less than twenty
ars ago. The radio map
rresponds to a length of only
out 10 light-years or less
an $1/10,000$ of the diameter of
e visible galaxy. The most
portant feature of the radio
ap is that the components are
igned with much larger
sible and radio features of
e galaxy – again implying a
croscopic 'memory' in the
urce of energy.

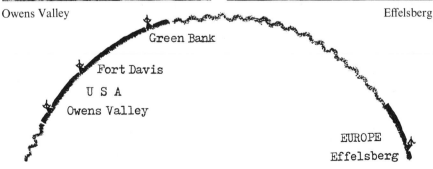

Green Bank

Fort Davis

U S A

Owens Valley

EUROPE
Effelsberg

155

reading a lot, and producing very little. They need not have worried. Hawking was learning about the behaviour of sub-atomic particles and readying himself for the next mental leap.

Although black holes might reveal their presence by very hot material falling into them, the black holes themselves had been thought of as cold, when regarded from the outside. They seemed perfectly, even sinisterly cold, because they absorbed any heat reaching them and gave nothing at all back to the universe. That was implicit in the very definition of a black hole, and in the rule that nothing could move outwards from inside the 'magic circle'. But Hawking found a loophole. The very intense gravity just outside the 'magic circle' could create durable particles. Some of them would promptly fall into the black hole, but others would escape.

Black holes were leaky. Hawking arrived at this conclusion reluctantly, because he shared the general conviction that any kind of escape from a black hole was impossible in principle. But there was no way round it, when he considered what happened in the vicinity of a black hole.

Hawking first invoked the creation of short-lived pairs of particles that would occur anywhere in empty space. Earlier (p. 26) we saw this process at work, with 'borrowed' energy being briefly converted into matching pairs of particles and anti-particles, which duly annihilated each other within the very short time allowed for their existence. But Hawking realised that the intense gravity near a black hole could intervene during the short life of particles and separate them, preventing their mutual annihilation. Most importantly, one member of a pair might fall into the black hole while the other escaped its clutches and travelled into the broad universe.

How a black hole can emit particles

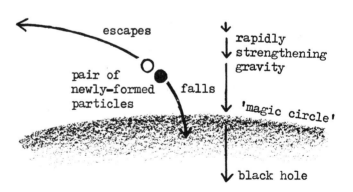

Although nothing was actually coming out through the 'magic circle', this process was equivalent to draining energy from the black hole's interior. Indeed another way of looking at the process was to see the energy of gravity serving to confer a durable existence on the otherwise shortlived particles by supplying them with 'real' energy. In principle any

156

kinds of particles could form in this way: electrons and protons, for example, and force-carrying particles of all kinds, including particles of light in the broadest sense. Hawking quickly realised that the release of particles by this subterfuge meant that black holes were warm.

Small black holes were hotter than big, massive black holes. The 'magic circles' of the latter were relatively large and the gravity at them was correspondingly weaker and less vigorous in creating particles. That made the warm black hole a very odd concept indeed. While a normal object became warmer if it absorbed energy and cooler if it released it, for the black hole the opposite was the case. When it gulped down matter or radiation from the outside universe it became more massive, its 'magic circle' enlarged and its temperature fell. When a black hole lost energy by its emission of particles, the 'magic circle' shrank and the temperature went up.

The idea of exploding black holes followed directly. An unreplenished black hole would lose energy and shrink. Becoming hotter it would lose energy more rapidly, thereby shrinking more rapidly and growing hotter still. The runaway process would culminate in the temperature shooting up to such a high level that the black hole was pumping vast numbers of particles and intense gamma-rays into the surrounding space. With this final outpouring, the black hole and the singularity at its centre would lose all their mass-energy and disappear from the universe.

A small black hole explodes and vanishes

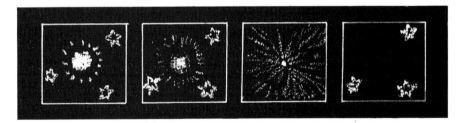

For 'conventional' black holes this fate would be extremely remote. The temperature of a black hole formed by the collapse of a heavy star was very low – less than a millionth of a degree above absolute cold. That was far cooler, in fact, than the radio energy filling all space with a temperature of three degrees, so that for the time being the black hole was absorbing energy from that source, and in its paradoxical fashion becoming cooler rather than hotter.

Ordinary 'stellar' black holes would not even begin slowly shifting towards an explosion until the whole universe had expanded and cooled very much more. Even then, their lifetimes were calculated with strings of 'times-tens' (10^{65} years) incomparably longer than the life of the universe so far. And for black holes a million times heavier, which might exist at the centre of galaxies, the lifetime would be a million million million times longer than for a stellar black hole (10^{83} years).

With deferments like those, the explosions of black holes might have seemed a mere footnote qualifying the earlier belief that black holes would last for ever. But Hawking was led into all the reasoning about explodability by a different speculation. He imagined that, at the birth of the universe, the Big Bang spawned large numbers of miniature black holes. Their lives would be much shorter and their explosions might therefore be observable.

Could black holes be exploding now?

Slightly uneven densities in the highly compressed matter and energy of the Big Bang could have produced black holes, Stephen Hawking reasoned. They would be black holes with all kinds of different masses, many much lighter and smaller than the black holes made by the later collapse of stars. Their sizes, temperatures and expected lifetimes could vary widely.

If the miniature black holes were scattered through the universe, the very smallest of them would have exploded long ago. Others would have lifetimes surpassing the age of the universe; in other words they were not yet due to expire. In between were black holes of a particular size that one might spot exploding in the twentieth century.

These were black holes no bigger than a proton, the very small nucleus of a hydrogen atom, but formed by the compression of about a billion tons of matter – say, the mass of a mountain. They would detonate with the violence of millions of H-bombs. As Hawking hastened to point out, the chances of it happening anywhere near the Earth were very small; by his reckoning the nearest probably lay well beyond the outermost planet, Pluto.

The principal sign, seen from afar, would be a brief, intense outburst of gamma-rays. They might be detectable from the ground, by the visible light created when the gamma-rays hit the Earth's atmosphere. And Hawking suggested that the new American Space Shuttle might be used to build a big gamma-ray detector in orbit about the Earth, to look for his miniature black holes exploding far away among the stars. Sudden bursts of X-rays from space had already been detected by American satellites launched to monitor the Nuclear Test Ban Treaty, but they did not match Hawking's specifications.

If miniature black holes existed in nature, as Hawking guessed they might, then the use of such objects as practical sources of energy became a degree less fanciful than it was in the early days of black-hole theory. It was still thoroughly far-fetched – if only in the literal sense that you would probably have to voyage a long way among the stars to find a suitable black hole.

Treating it more warily than you would handle a deadly virus, you might then attempt to electrify it and cage it in an extremely strong electric force. Shifting a mass as great as Mount Everest would not be easy but, if you were clever, you might make it propel itself. In principle you might nudge it into orbit near – but not too near – the Earth. There it could solve human problems of energy supply for all time. By feeding the captive black hole with matter of any kind (for example, noxious radioactive wastes) you could generate as much energy as you wished and beam it to the Earth. Other, more advanced civilisations elsewhere in the universe might already be employing such devices.

It was all very thought-provoking. But some colleagues found strong reasons to question Hawking's optimism about detecting the explosions of miniature black holes, and indeed about whether they existed in any significant numbers. The most favourable estimates of their production in the Big Bang came from a theory which did not accord with the trends and discoveries of high-energy research; the so-called 'bootstrap' theory (p. 93). It said that there were no truly fundamental particles. Very high temperatures, as in the Big Bang, would create more and more types of massive particles and promote the production of miniature black holes.

The success of the quark theory made bootstrap unfashionable, to put it mildly. For black holes to form in large numbers in a 'quark soup' during the Big Bang required much larger fluctuations in density than the boot-strap theory needed. It was hard to see where those fluctuations would come from. The rival particle theories also painted very different pictures of what happened at the high temperatures of a black-hole explosion. A 'bootstrap' black hole would flash a hundred thousand times more in-tensely in its final stage than a 'quarky' black hole. In short, the dominant theory of particle physics made both nature's creation of miniature black holes, and their detection by humans, seem more difficult.

On the other hand, that added spice to the search. Observing an exploding black hole would not only be an extraordinary test of the theory of gravity. It would be an exceptional experiment for high-energy research, capable in principle of deciding once and for all between the rival theories. In addition, it could give fresh insight into the nature of the Big Bang. Even a failure to find any exploding black holes might be counted useful information. It would mean that the Big Bang was a very smooth process – too tidy to produce the small black holes.

In any case, the importance of Hawking's idea did not depend on the actual existence of miniature black holes. Even while he was heaping speculations upon guesses upon hypotheses, his reasoning about what *could* happen in extreme gravity exposed deep connections implicit in the universe and its contents. Those connections could stand even if, for historical reasons, nature had made no black holes of any kind.

How nature abolishes history

Some of the bold ideas developed by Stephen Hawking, while playing the game of unifying the laws of nature, at the very edge of a black hole, had already occurred to other people. In particular Jacob Beckenstein of Princeton saw a similarity between the laws of black holes and the laws of heat. He stopped short of Hawking's crucial step of explaining how black holes could be warm.

Twentieth-century science had already made nonsense of the peda-gogues' fussy compartments of thought. It brought higher mathematics into embryology, for instance, and used X-rays to illuminate genetics. Hawking managed to put 'gravity', 'elementary particles' and 'heat', formerly quite different topics of study, into a blend of new understand-ing. By relating the laws of heat to black holes, he found an unlikely connection between a cosmic force (gravity) and human notions of disorder and information.

Heat is the disorderly motion or vibration of the small constituents of matter, whether molecules, atoms or particles. It is, of course, necessary for light and for life. The catch is that you cannot exploit heat for any useful purpose without paying a tax on it, or throwing sops to the dragon of disorder which continually threatens to abolish all interesting objects in the universe. For example, a growing tree or a human body organises atoms into very special combinations prescribed by information in the genes, establishing order out of the disorder of the soil or the digested food. These are wasteful processes, in the sense that they exude a lot of low-grade heat into the surroundings. Thus they increase the random motion elsewhere. In the end you might expect the universe to finish up with the meaningless uniformity of a lukewarm bath.

The laws of heat cover everything from the growth of a tree and the running of a car engine to the behaviour of particles and energy during the Big Bang. They prohibit any kind of perpetual-motion machine in the universe or on Earth. The first law says you can't get something for nothing. The second law says you can't break even, because disorder in the form of random motion must, on balance, increase.

160

Black holes, Hawking found, were warm within the strictest meaning of the laws of heat. The nature, variety, energy and rate of the particles they emitted corresponded exactly with those from a conventional hot object. Hawking could therefore assign a precise temperature to a black hole, and it depended simply on the surface gravity of the black hole, at the 'magic circle'.

All black holes of the same mass, however formed, had the same temperature. Corrections were needed, according to any spin and electric charge that the black hole might possess, but again the history of how it obtained those properties made no difference. The mass, surface area and temperature of a non-rotating black hole were fixed: a certain mass, or a certain surface area, *equalled* a certain temperature. Hawking's equations creating the connection between gravity and heat were among the most extraordinary ever written.

The first law ('You can't get something for nothing') also applied to black holes, if one specified the mass-energy that nature had to expend to make a black hole bigger, or to make it spin faster. But the parallel with the second law ('You can't break even') was especially striking. The surface area of a black hole, at its 'magic circle', was a direct and absolute measure of the degree of disorder in it. By swallowing more matter, the black hole increased its surface area, thus increasing this measure of disorder. In the explosion of a black hole the surface area would diminish, but the randomly emitted particles represented a very high degree of disorder.

Disorder, in the sense of the laws of heat, was fully equivalent to a loss of information. There was no way of recalling the shape of a snowman once it had melted. Nor was there any way of deducing the history and contents of a black hole. Jacob Beckenstein had first suggested that the surface area was a measure of all the possible ways in which a black hole might have been made, and hence of the degree of disorder which it represented. In Hawking's scheme too, black holes destroyed information. What came out of an exploding black hole bore no relation to what went into it. Whatever way you looked at it, the dragon of disorder inhabited the black holes. It abolished history, destroying information about the pieces of the universe that it devoured.

Most of Hawking's fellow theorists were content to welcome his results, to note that black holes fitted amazingly well with the ideas about heat, and to go on to explore the implications of the Hawking Process – the exploding black holes. While this industry boomed, Hawking himself brooded about the meaning of the parallel between black holes and disorder, and how black holes might set limits to the human ability to know the universe and predict its behaviour.

He imagined extremely small black holes being created briefly, but anywhere and all the time, out of 'borrowed' energy. They could nibble away at information and order on the sub-atomic level. And at the other end of the cosmic scale very distant galaxies, moving away too fast for their light ever to reach us, might be aspects of the same theme of disorder and loss of information in the universe. Hawking considered that it added up to a new kind of uncertainty in the human perception of nature.

While fellow-theorists were expressing doubts and even irritation about that idea, Hawking was broaching an even more controversial one, suggesting that the laws of nature might appear different to different observers. It threatened to stab Einstein's theory in the heart. Had Hawking overstepped the bounds of useful craziness? When someone like him was juggling with ideas the onlooker could only hold his breath.

Gravity remains detached

The demolition of matter in a black hole broke once-cherished laws of the particle physicists – that the total number of heavy particles and the total number of electrons in the universe could not change. But those laws were already crumbling. The idea that quarks and electrons were related led to the possibility of protons very slowly decaying into anti-electrons, which could go on and annihilate electrons. It remained an open question whether there might be some deep-lying link between permissible annihilation by that means, and annihilation in black holes.

For the general enterprise of trying to make sense of particles and forces, Stephen Hawking's theory of how very intense gravity at the edge of the black hole could manufacture particles was of special interest. He had more success than most of the people who had been trying for half a century to reconcile particle physics and gravity. But it was still only one bite at a larger problem.

Quarks and electrons were known to be very small indeed. Could gravity fashion them into something like the 'singularity' envisaged at the heart of a black hole, where all the matter of a star was collapsed to a single point? Another issue for the theorists concerned Einstein's success in relating gravity to the basic concepts of space and time. Could the other forces and the particles on which they acted also be married to the character of space-time itself? As Abdus Salam put it: was space-time grainy and was that where the qualities of the particle came from?

Gravity operated effectively over the vast distances of the universe. In the micro-universe it became completely overshadowed by the strong nuclear force, the electric force and even by the weak force. But that could

162

be only an interlude, as it were. When particles approached very closely together, gravity ought to become relatively strong again. The sheer proximity of particles should enable gravity to assert itself powerfully, much as it seemed to do in a black hole.

At extremely short distances between particles the theories of forces would inevitably crack. Einstein's theory of gravity itself predicted an infinite force – possibly the same absurdity that dogged the calculations of other forces for many years. And what really happened when gravity ran up against the quantum theory, with its notions of 'borrowed' energy? Direct attempts to string a rope between the two theoretical peaks of Einstein's gravity and the quantum theory had failed repeatedly. Yet this was the biggest single challenge for those searching for the key to the universe. To some, it seemed that only rebuilding both mountains would serve. Roger Penrose of Oxford, one of the architects of black-hole theory, sought a desperate remedy of that kind.

Penrose set out on the long and arduous task of trying to redescribe the universe and its contents in terms of 'twistors'. These were curious creatures of the imagination (objects or particles would be too strong a word) whose only properties were spin, energy and a capacity to travel about at the speed of light. Out of them, Penrose hoped to reconstruct the cosmic forces and the particles without appealing to established ideas about space and time. He was marrying geometry and the behaviour of particles in the very concept of the twistor itself.

Others continued, with less extreme policies than Penrose's, the efforts to bring about the wished-for unification. Particle physicists were generally inclined to concentrate on the real progress they were making, and leave the links with gravity for their successors to worry about. Sheldon Glashow remarked in 1975:

'We agree to abandon all hope, for the moment, to include gravity in the theory. Although an ultimate synthesis must correct this omission, we assume that a reasonably coherent model of particle physics without gravity exists, and that it is accurately predictive at sufficiently low energy.'

At just what energy would gravity become important in particle physics? The general view, based on Einstein's theory, had it at more than a billion times the energy of any accelerators contemplated in the 1970s. Others, notably Abdus Salam, suggested that new gravity-like force-carrying particles might come into play at considerably lower energies. Either way, experiments in sub-nuclear gravity lay in an unknown future.

The notions of gravity and mass were inextricably tied together, because gravity gave expression to mass. The gauge theories of particle

163

physics, even amid the triumphs of the 1970s, made no progress in explaining why a quark was heavier than an electron, or even why a heavy electron was more massive than an ordinary electron. Richard Feynman, one of the ablest of theorists, retained a reserved scepticism about the whole gauge-theory excitement because, as he put it, the problem of the masses of the particles had been swept into a corner. Perhaps the solution had to wait for a deeper understanding of the gravity connection.

Whatever might go on in black holes, large, small or ultra-small, it was mild compared with the close confrontation of all the constituents and forces of the universe at the supposed moment of its creation in the Big Bang. If the key to the universe was a matter of explaining why the cosmos and its contents had to be the way they were, you could take stock of the state of human knowledge by seeing how good an account it could give of the origins. That is the theme of my final chapter.

is is the actual
culated size of a black
le of the same mass as
Earth. If you were
0 kilometres away
m it, the gravitational
ce would feel the
ne as at the Earth's
face. But at 60
timetres (two feet)
m it you would
perience a force 100
llion million times
onger.

6 Universe-maker

New generations of human beings will no doubt despise the twentieth century for its avoidable poverty, disease and war. They may nevertheless be envious of a period of history when there were still fundamental secrets that due efforts of will, imagination and diligent experiment could wrest from nature; when people working with boyish enthusiasm and unfashionable optimism could make an indelible mark on human knowledge.

The speculations encompassed, of course, other, possibly superior, intelligences at work on the planets of other stars. But for all practical purposes, the physicists and astronomers of the twentieth century were the first agents of the physical universe through whose eyes that universe could look with a wild surmise upon its own character and destiny. As they explored the realms of the very large and the very small they found all the paths, all the timelines, converging in the first tumultuous moment when the universe came into being: in the Big Bang.

Reddened galaxies and warm sky

Astronomers have benefited like everyone else from the gadgets of electronics. When I visited observatories in the late 1960s, they still squatted uncomfortably the whole night through, in the midst of their great optical telescopes. They might spend several hours tracking a single object in the sky, trying to gather enough light to register on a photographic plate. By the mid-1970s, the electronic revolution had come to optical astronomy.

Very sensitive detectors, capable of responding even to a single particle of light, stood where the astronomer once put his eye. They greatly increased the range, power and sharpness of gaze of the big telescopes. The astronomers were gathering the information they wanted about many more objects, in the course of one night. They sat in comparative comfort in an observing gallery, seeing the universe by television, as it were, and converting their faint specks of light into a form that computers could analyse.

Their work was directed at clarifying the history of a universe that had clearly evolved. Until the demise of the so-called Steady State Theory there had been room for doubt on that score, but not any longer. The

dividual particles of light corded in the image-ensifying television system, veloped for astronomy by ec Boksenberg of University ollege London. The raw gnals shown here go through ectronic processing to give ry rapid and precise formation about distant jects in the universe. The stem has been installed in major optical telescopes and s helped, for example, to entangle confusing aracteristics in the intense ht emitted by quasars.

astronomers were therefore like archaeologists of the cosmos. Because of the peculiar effect of the travel-time of light, astronomers saw the universe almost inside out, with the earliest and most crowded galaxies and quasars lying in the farthest shells of observable space. The earliest event of all, the Big Bang, was spread all around.

Some radio astronomers thought they had already reached right back to the era when galaxies were first being formed, beyond which the sky became empty.

With their sensitive light detectors the optical astronomers were striving to identify and measure the distances of as many as possible of the distant quasars, to judge if that impression was correct; also to see whether there was a period, perhaps in the youth of the galaxies, when great explosions were more common.

Half a century earlier, the idea that the universe itself had exploded seemed like one of the crazy ones. Yet in whatever direction Edwin Hubble turned his telescope, from Mount Wilson near Los Angeles, he saw distant galaxies moving away like shrapnel flying from a bomb. Recognisable light from particular atoms in the galaxies had lost energy and was shifted towards red; the simplest interpretation was that they were flying away at high speed. The farther the galaxies were, the greater was the reddening and the faster their receding motions.

When Hubble traced their motions back, the galaxies all seemed to have started from the same place. By 1927 Georges Lemaître, a Belgian priest, was proclaiming the birth of the universe by disgorgement out of a primeval mass. There were difficulties, though, about time-scales, and the physics of the day could scarcely begin to cope with the behaviour of matter in that primeval mass.

The modern account of the likely birth of the universe in the Big Bang began with work, from 1946 onwards, by George Gamow, a Russian-born theorist who settled in the United States. It appealed to his sense of fun that he could join forces in 1948 with colleagues named Alpher and Bethe, add his own name to theirs, and develop the Alpher-Bethe-Gamow theory of the Big Bang, grossly punning with the first three letters of the Greek alphabet. It did not win immortality, though; some important ideas from 'Alpha-Beta-Gamma' survived, but the later descriptions of the Big Bang brought essential additions and corrections.

Hot Flash might have been a slightly more appropriate name than Big Bang. The newborn universe was, by any reasoning, dominated by very intense rays. Gamow realised that some energy of the radiation must have been left over. Empty space throughout the wide universe should not be completely cold. In 1965 came the chance discovery, from the Bell Tele-

side-out universe. The sheer
time required for light to reach
the Earth from distant parts of
the universe enables
astronomers to see it as it was
millions of years ago. But the
effect of this is to put the oldest
and most compact eras of the
expanding universe at the most
extended distances all around,
and the most recent, dispersed
as close at hand.

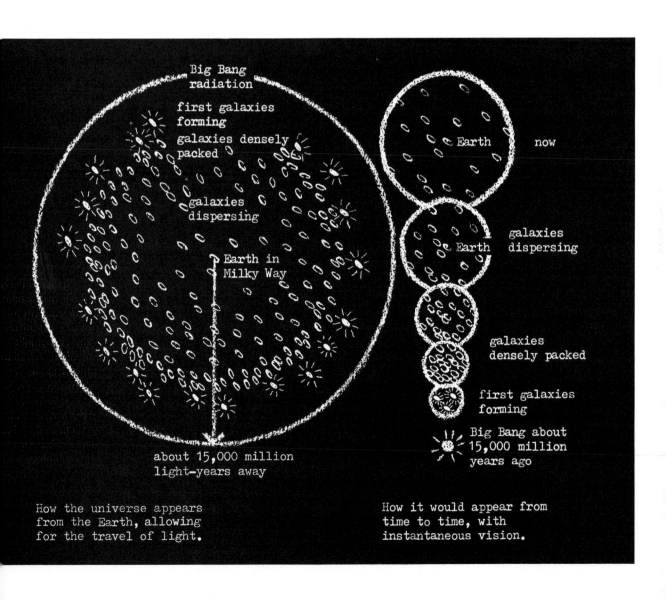

Big Bang radiation

first galaxies forming

galaxies densely packed

galaxies dispersing

Earth in Milky Way

about 15,000 million light-years away

Earth   now

galaxies dispersing

Earth

galaxies densely packed

first galaxies forming

Big Bang about 15,000 million years ago

How the universe appears from the Earth, allowing for the travel of light.

How it would appear from time to time, with instantaneous vision.

Taking the temperature of the sky. Radio energy fills space, produced by the cooling of the flash of the Big Bang. The fact that it is more or less uniform in all directions does not mean that the Earth is at the very centre of the universe – although by definition it is at the centre of the observable universe. Small differences in the apparent temperature of the sky – the energy of the microwave radio background – are expected because of the local motions of the Earth. Two of them are shown here – the motion of the Earth around the Sun, and the much faster motion of the Sun (+ Earth) around the centre of the Milky Way (speeds in kilometres per second). The Sun is travelling towards the constellation of Virgo and away from the constellation of Aries. That should produce a warming by about two-thousandths of a degree in the sky in Virgo, a corresponding cooling of the sky in Aries, and lesser changes elsewhere. A U2 aircraft (below) has been fitted with microwave horns to compare sky temperatures in different directions.

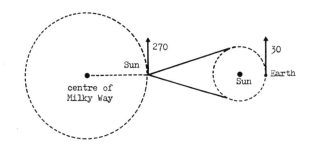

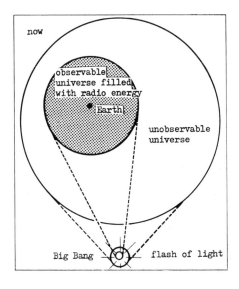

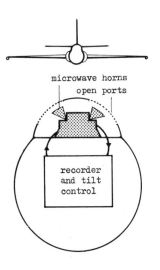

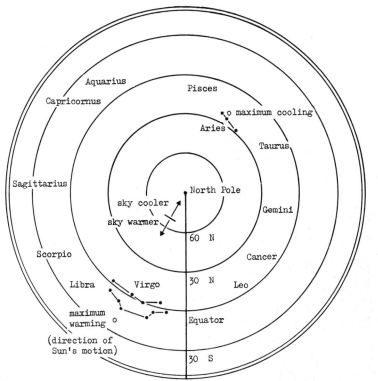

phone Laboratories in New Jersey, that radio energy was apparently pervading all space with a temperature of three degrees above absolute zero. The sky was warm. It provided exceptionally strong support for the idea of the Big Bang.

The impression that the universe was still 'inside' the bath of the Big Bang was not misleading. It was not an event that happened 'somewhere else' as one might behold an exploding volcano through a telescope. The stuff of the Milky Way and the Earth originated inside the explosion and the universe expanded all round our piece of it.

After the discovery of the universal radio energy – the 'microwave background' as it was called – measurements showed that it was remarkably uniform in all directions. The universe as a whole was neither lopsided nor very lumpy. That made calculations of its early history much easier. But it was unthinkable that the Milky Way and the Earth were at rest in relation to the universe as a whole, and by the mid-1970s further experiments were trying harder to find slight differences in the warmth of the sky. In one experiment, a high-flying U2 aircraft of the National Aeronautics and Space Administration was making nightly flights from an airfield in California, fitted with special radio equipment from the Lawrence Berkeley Laboratory.

Any movement of the Earth relative to the universe as a whole should show up as a definite warming of the sky in the direction of motion through the 'sea' of radio energy. The Earth was swinging around the Sun once a year. The Sun was swinging around the centre of the Milky Way, at 270 kilometres a second. The odds were very long indeed against the Milky Way being stationary. The Earth had to have *some* motion relative to the universe. A failure by the U2 or other ground-based experiments to find any motion would create a great headache for the theorists.

A futile future for the universe?

Every galaxy felt the gravitational pull of the other galaxies, and of all the contents of the universe. Accordingly the speed at which they flew apart steadily diminished with the passage of time, much as a rocket was slowed down by the pull of the Earth behind it. After Einstein's theory of gravity and Hubble's discovery of the recession of the galaxies put the study of the overall nature of the universe on its modern course, a question soon arose. Was gravity strong enough to overpower the galaxies and drag them back together again? Or would they go on flying apart, like space-rockets escaping completely from the Earth?

In the 'closed' version, the eventual outcome would be a decisive collapse into an unimaginable holocaust, like the Big Bang run backwards. Many people fancied this picture. Although time would come to an end as far as our universe was concerned, the aggregation of matter and energy suggested the possibility of a new universe emerging from the demolition of the old one – and so on, in a cycle of universes, perhaps indefinitely. In an 'open' universe, our package of space and time would persist, with the galaxies travelling farther apart for ever.

In the mathematics of the cosmos, the choice between an open or closed universe was a fairly casual one, depending on the average density of matter. The issue was not unimportant for anyone trying to understand the universe and its contents. Knowing the answer would allow you to estimate the amount of invisible matter. It might take the form of black holes or very diffuse gas or perhaps, as some theorists were beginning to suggest, neutrinos with non-zero mass. It was a desirable piece of information which could shed light on the nature of the Big Bang.

An open universe would testify to the violence and effectiveness of the Big Bang. A closed universe, by the associated idea of our universe being born phoenix-like from the collapsed ashes of its predecessor, could conceivably help to solve some, but by no means all, of the riddles about how the Big Bang came about. Yet most of the questions were technical rather than philosophical.

Evidence in favour of an eventual collapse of the universe looked fairly strong in the late 1960s. At least one distinguished astronomer, Allan Sandage of Palomar, was measuring the speeds of the galaxies and saying that they would halt in their tracks in about thirty billion years; the heavens would then fall. Other observers were detecting gas in unexpected quantities in the spaces between the galaxies, which might add sufficiently to the gravity of the universe to make such a collapse inevitable. The evidence soon shifted strongly – though not decisively – in favour of an 'open' universe.

Just how the vacillating answers from the astronomers might settle down was an interesting but hardly, you would think, an earth-shaking question. It made no difference to life on our planet, which was doomed in any case by the remote but eventual exhaustion of the Sun. If you cared to imagine our own descendants or other intelligences striving to survive by moving from star to star or galaxy to galaxy, even for them the question could not assume practical importance for many billions of years.

And yet the issue of whether or not the universe would go on expanding for ever touched nerves of religious sensitivity. That brought another element than science into discussions about the rate at which the galaxies were slowing down. The closed and oscillating universe reminded many

sed'
verse

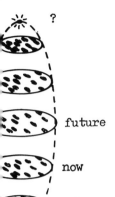

? future now past Big Bang

en'
iverse

ntinues expanding
ever .

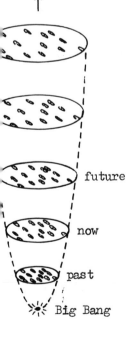

future now past Big Bang

people of the endless cycles of birth and rebirth envisaged in the oriental religions. Materialists were inclined to prefer that picture to a singular event of creation. Marxists had, from Engels, an explicit insistence on an eternal succession of worlds. On the other hand the sudden creation, apparently out of nothing, was strongly favoured by the Catholic Church and the Pope officially adopted the Big Bang theory in 1951.

After Galileo's time, however, cosmological issues were settled as far as possible by observations and measurements, not by religious or aesthetic preferences. By the 1970s industrious and open-minded astronomers were finding that completely different approaches were tending towards the same answer: the universe was open and would go on expanding for ever.

For a start, Sandage had changed his mind. He and his colleagues, notably James Gunn, were amassing measurements that suggested that the rate at which the galaxies were slowing down was not, after all, very great. Unfortunately the fact that galaxies would have changed their appearance with the passage of time, by evolving and perhaps gobbling up their neighbours, made estimates of their distances unreliable.

The amount of heavy hydrogen in the universe was a second guide to whether there was enough matter in the universe to bring about its eventual collapse. The Big Bang evidently manufactured helium gas, in quantities that did not depend critically on the density of the universe. The residue of heavy hydrogen was more sensitive. In a very dense universe virtually all the heavy hydrogen would have had the opportunity to build up to make helium during the Big Bang. But there was a lot of heavy hydrogen about. It was present in tap water and in the spaces between the stars as well. Again there were uncertainties, both in measurements of heavy hydrogen in the universe and in the theories of its production in the Big Bang. But the indications clearly favoured a less dense, and hence an open universe.

Other astronomers tried to work out the slowing down of the universe by estimating its true age, from the calculated ages of stars and chemical elements. While all of these methods were attended by great uncertainties, when put together they pointed to a universal density too small, by a factor of about ten, to halt the galaxies in their tracks. But anyone wanting a collapse, for scientific or religious reasons, might still claim that the possibility remained alive, pending further measurements.

An interesting technique that could in principle check the density and fate of cosmic matter involved measuring the curvature of space in the universe. The curvature and the consequent bending of the paths of light or radio energy could make the universe act like a great lens.

Identical galaxies seen at increasing distances would look smaller and smaller until, beyond a certain distance, they would look bigger again, because of the gravitational lens effect. The denser the universe, the sooner this enlargement would become apparent. But, in practice, the usual questions arose, about galaxies and how they might change their character and size with the passage of time.

If you accepted the prevailing opinion that the universe was 'open', the end of the universe looked excruciatingly slow. New stars would go on forming and burning and expiring in the galaxies; new planets too, with perhaps new life forming in the rays of long-lived stars. But gradually, inexorably the galaxies' resources of gas would be exhausted. As stars exploded or faded away without replacement, the splendour of the galaxies would be dimmed.

Add enough zeros to the age of the universe, and it would become almost wholly black, except for those increasingly distant smudges of light that might tell some surviving astronomer on another planet what galaxies were. Add more zeros, and the universe would consist only of dead stars and black holes in skeletal galaxies, still hurtling through space. The very small chance of a star remnant blundering into a black hole might become near certainty, given enough time. But the action would not be over: the cosmic forces would have to play the game to the end.

The weak force, remembering the quark's essential kinship with electrons, would very, very slowly transmute all the protons into anti-electrons and neutrinos, decomposing all the atoms. The anti-electrons would maintain the scrupulous balance of electric charge in the universe, and would in principle annihilate with the electrons made redundant by the disappearance of the protons. But with the matter of the universe extremely dispersed, many anti-electrons would escape annihilation and parade for aeons through the darkness, as ghosts of the protons.

The second process would be more dramatic if there were anyone around to witness it. The darkness would be broken by colossal explosions, when the time fell due for massive black holes to expire. New protons produced in those explosions would scatter unconstructively, surviving for yet more aeons, until they too decayed into anti-electrons. The cosmic forces would thus be left acting out a futile rigmarole.

It seemed strange that the creation of conditions in which life could exist, here and there for perhaps 100 billion years, should imply so much unfinished business afterwards. It would drag on through an utterly unimaginable desert of time, clocking off as many years as there were atoms in the universe, and then starting the count again. The unfashionable theory of the collapsing universe had the merit of a tidier doomsday that you could write in a conceivable calendar.

Speculations about alternative universes

A human being can jot down a prescription for a universe quite different from ours, oh the back of an envelope. That is, perhaps, a sign that the human imagination is bigger than the universe. It may be also much too easy an exercise to have much meaning. The ancient Greeks were at it, drawing universes in the sand, and we know that they were wholly ignorant of most of the forces and qualities of matter that later seemed indispensable for understanding how a universe was built. Without spoiling the fun, we ought to keep in mind the legitimate criticisms of such casual universe-building if only to conserve envelopes.

Other universes are unobservable; if you could see them they would be parts of our own universe. Nevertheless, for twentieth-century physicists as for the ancient Greeks, alternative universes have provided a possibly useful way of thinking about the universe. They may also exist, if you do not define 'exist' too strictly. Conceivably you might even infer the existence and nature of another domain of space and time, from subtle features of our own. There have been serious speculations about a co-existent anti-universe, or other realms beyond the horizon where the laws of nature were not the same. Is our universe a black hole in somebody else's universe? Can the supposed singularity at the heart of an ordinary black hole constitute a tunnel into quite different domains of space and time? It is often hard to tell whether the proposers of such concepts have their tongues in their cheeks.

In some respects, the character of our universe is uncanny. It seems to be perfectly adjusted to enable life and intelligence like ours to evolve and prosper in comparative safety for very long periods. A denser universe might have quickly collapsed together into a fiery mass, extinguishing all life. Even without that extreme catastrophe, life on Earth depended on the steady burning of the parent star. It took 4500 million years of planetary and biological evolution on Earth to create microbes and let them change successively into worms, fishes, reptiles, mammals and eventually humans. If the particles or the cosmic forces were slightly different, the Sun could have expired already.

If quarks were lighter in weight, or electrons were heavier, stars would burn faster. Again, a quantity called the 'fine-structure constant' defines the strength of the electric force, or the probability at any moment of finding a force-carrying particle of light in the vicinity of an electron. Its value is roughly $1/137$ but, in calculating the brightness of a star, you have to multiply the 'fine-structure constant' by itself no less than twenty times. As a result, a small change in the figure would produce enormous differences in the character of an ordinary star like the Sun, making it either much dimmer or much more fierce.

Thoughts of this kind led an astro-theorist, Brandon Carter, then of Cambridge, to propound the Anthropic Principle for inhabitable universes. It represented a reaction against the Copernican Principle that ran through astronomical thought after the Earth was rudely dislodged from the central position it occupied in the medieval universe. The Copernican Principle said that the Earth occupied no privileged position; for example, one could take the sample of galaxies observed most readily from the Earth as typical of the universe as a whole.

The men who conceived the Steady State Theory of the universe invoked the Copernican Principle strongly. They asserted that the universe was much the same anywhere and at all times – here and now, or a million million light-years away. But the downfall of the Steady State Theory, and the evidence that the universe looked different at great distances, weakened the Copernican Principle. It remained a useful corrective to any tendency to think too highly of the human position in the cosmic scheme.

Carter declared that our situation in the universe *must* be privileged, to some extent. For a universe to be observable it had to be inhabitable and that might be possible only at particular places and in certain periods. Observers could not survive, presumably, in the vicinity of an X-ray star, or at a time before an adequate supply of heavy chemical elements was distributed by the explosions of early generations of stars. Such was the essence of the Anthropic Principle. It implied, for instance, that an observable universe has to be at least some thousands of millions of years old.

Some theorists rejected this line of argument. They scorned excessive brooding about the conditions needed for life – the precise strength of the forces, the supply of chemical elements, or the time needed to put them together into an intelligence capable of gazing upon the universe. Life could, they said, adopt forms wholly different from anything we knew or even imagined, and might evolve much more quickly.

In its strongest form, the Anthropic Principle noted that the overall character of an observable universe had to be suitable for the creation of observers. That suggested the possibility of other kinds of universes coming into being, and existing for aeons or for ever, and fashioning weird inanimate objects, without any inhabitant being aware of its existence.

Imagine a set of millions of different universes – a 'world ensemble' was Carter's phrase – possessing between them every conceivable combination of starting conditions and strengths of the forces. Then a sort of natural selection could occur, whereby some, but only some, of the universes would support intelligent life, forming a 'cognisable subset' of universes. In them, the strengths of the various cosmic forces would fall within certain ranges and combinations, of which our 'cognisable' universe presented a typical sample.

John Wheeler of Princeton took this argument one stage further. If the vast majority of universes were sterile, because the sizes or forces were wrong, it would be extravagant or at least improbable for nature to overshoot and create conditions in which many different forms of life were possible in many different places in one universe. Many astronomers supposed that there must be many inhabited planets circling around many stars, with other intelligent beings on them, perhaps already busy communicating by pan-galactic radio networks. Wheeler speculated otherwise: if life came in only by a 'narrow squeak', perhaps the odds were that the Earth was the only outpost of life in this universe.

Whatever one might think of those arguments about other conceivable universes, as attempts to describe and explain nature, they at least helped to illustrate how firmly the character of life on Earth was embedded in cosmic laws; and that our universe was above all a liveable universe, a suitable home for human beings. To the extent that we were part and parcel of the cosmos we might learn to look upon the fiery stars or exploding galaxies without horror, as alternative expressions of the friendly cosmic forces at work.

I should add that Carter himself regarded the 'natural selection of universes' as an argument of last resort. He, like all his colleagues, would much prefer to find that the size, history, particles and forces of the universe were fixed for some more logical reason. In other words the preference would always be for explanations that left no latitude for varying the forces and constituents of the universe, so that any imaginable universe might be very like our own.

Broken symmetries

The problem of finding some logic to the cosmic forces and particles revolved around the issue of the broken symmetries. We encountered it, you may recall, in connection with the theory that unified the electric force with the weak force. Theorists wondered why the force-carrying particles of the weak force, the W particles, were very heavy, while particles of light had no mass. It was a discouraging broken symmetry. But it made more sense when ghostly particles were invoked which weighed the W particles down, but not the particles of light.

By that breaking, the universe was armed with additional forces, which helped to make human existence possible. If any single concept seemed fit to be an emblem for natural philosophy towards the end of the twentieth century, it was the broken symmetry. It signified above all that the universe was a compromise between simplicity and complexity, order and variability.

The philosophers of ancient times wondered fancifully how the universe might be built. Parmenides, who lived at Elea some twenty-four centuries ago, disliked the notion of empty space. He declared that the universe was filled, everywhere. Any unevenness in the universe would imply an emptiness at one place compared with another, which Parmenides ruled out on principle; any motion, too, would have a similar effect. Parmenides' universe therefore was like a ball of glass, perfectly spherical and uniform, and incapable of including any movement or change, any beginning or any end. The fact that the world appeared to be a bustling place, full of movement, birth and decay, was to Parmenides just an illusion.

At the other extreme was the somewhat older theory of Anaximander and Heraclitus, who lived in Miletus and Ephesus respectively. For them everything in the universe was in flux – in perpetual movement and change. Some basic stuff (Heraclitus took it to be fire) was continually exchanged with familiar matter. The observable, apparently durable features of the world were illusory – in a less extreme way than Parmenides would have it, but in the sense that a river appeared the same on different days although the water it contained was all the time being renewed.

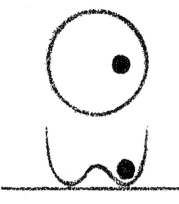

How two symmetrical objects
can create an asymmetrical
outcome. As Steven Weinberg
points out, a ball in a bowl
with a central dimple is obliged
to break the symmetry by
adopting an off-centre
position.

Those contrasting ideas, of total rigidity or total flux, were both offensive to common sense. Yet applied to particular features of the universe, rather than to the universe as a whole, they could be useful. When you declared, for instance, that the law of gravity was always the same everywhere you were making a 'rigid' statement like Parmenides. Equally, if you supposed that the constituents of a real gas (air, for instance) were in random internal motion you were thinking in the 'fluxive' manner of Heraclitus and you were not deceived.

The atomic theory, as framed by Leucippus of Miletus and Democritus of Abdera, was an early compromise between the extremist philosophers. The principle of flux was preserved in the atomic theory; so was the rigid sphere of Parmenides, but it was cut down, from the size of the universe to the size of an atom, too small to be seen. What was in flux was a vast number of such atoms, each possessing the Parmenidean qualities of rigidity and infinite durability. Nineteenth-century chemistry came back to something very like the Greek atomic theory.

178

With hindsight, the 'fire' of Heraclitus was not too different from the 'energy' from which everything was made, according to twentieth-century physics. But beware of giving too much credit to the Greeks. They thought of so many possible universes that it is scarcely surprising that some of their speculative philosophy turned out to be recognisable, in the conclusions of the incomparably more powerful experimental philosophy of modern science.

In any case, the Greeks and the nineteenth-century chemists missed the crucial point that was to become clear in the twentieth century: Parmenidean perfection was no use for atoms, either. They would not work. Indeed, the most nearly perfect atoms, such as helium and krypton, constituted almost completely inert gases.

Putting it in modern terms: if all the basic particles of nature, the quarks and electrons, were just one kind of particle, and if all the cosmic forces were just one kind of force, the universe would have a certain elegance and purity, but it would be hopelessly boring, like Parmenides' glass ball. Nothing could change, nothing could live. If, at the other extreme, there were too many basic particles and forces, and too little uniformity, then nothing could be durable, nothing could evolve in a systematic way. The perfect symmetry had to be broken, but not extravagantly.

Among all the art and architecture that embodied the human love of symmetry, the traditional patterns of the Moslem world most aptly captured the delicate notion of the broken symmetry. The flowerbeds of Moghul India confined living plants, in all their natural variability, within rigorously geometrical outlines that represented the order and stability of the universe, or of God.

The same symbolism appeared in a million impersonal patterns that decorated mosques, harems and carpets throughout Islam. Repetitive symmetrical patterns enclosed emblems of life: leaves, fruits or animals. The geometry was both a cage and a stage for life. Bridging any conceptual gap between order and life were the holy words written in stone, because an alphabet combined the regularity of geometry with the fluidity of life.

But broken symmetries abounded in any earthly scene: the shadow on a face; the handle on a cup; the spin on a ball during a game; the petals of a flower which, if they were too regular, we should suspect were made of plastic. Controlled imperfection was the hallmark of reality and, as it transpired, of the universe.

The long human search for hidden perfection, as a wished-for feature of the key to the universe, was striking less because of its persistence (which could have been boneheaded) than because of its approximate success. The human mind would accept perfection in cosmic matters as almost self-explanatory, but there was no real guarantee that nature would oblige. Time and again, discoveries rudely dislodged perfection, from the Moon and the stars, from the atoms and the particles. Yet each time the outward appearances became tarnished, there were new laws that made up for it.

In nature, a broken symmetry was not a casual piece of untidiness, or a clumsy artist's lopsided effort. It implied a perfect symmetry which was then violated in a precise way. The breaking occurred according to laws just as strict as those which established the symmetry in the first place. The perfect symmetry was not destroyed but hidden. By reversing the processes that broke it you could, conceptually at least, recover something like perfection.

With the advances in knowledge of the cosmic forces it became possible to imagine circumstances in which all the cosmic forces would be more or less equal. Then all particles of matter would behave in much the same way, and there would be something like perfection again. For a start, the weak force would become relatively stronger in hot conditions. At a high enough temperature it would become equal in strength and range to the electric force. The W particles carrying the weak force would lose their heaviness, and the true kinship of the forces would appear.

The colour force, which bound the quarks together and gave rise to the strong nuclear force as a by-product, was peculiar. It was strong at comparatively long ranges but became weaker when quarks approached close to each other, giving them the 'freedom' we encountered earlier (p. 89). So you could envisage conditions in which quarks were always so tightly packed that the colour force between them was no stronger than the electric force. Finally, if quarks and other manifestations of energy were crowded together even more tightly, the force of gravity between them would become just as strong as the other forces.

In short, by strengthening the weak force and gravity, and keeping the colour force slack, they could all finish up with about the same strength as the electric force. The necessary circumstances for that to occur naturally involved a very dense packing of matter and a very high temperature. And where would such a 'quark soup' be found? In the Big Bang.

The first split second

The Big Bang was such an awesome event to contemplate, being all-embracing in its massiveness, violence and creativity, that physicists might have been forgiven shrugging it off as a kind of impenetrable chaos.

Yet, confident in the strong grip that the twentieth-century mind had on the physical world, human beings attempted by straightforward reasoning to encompass, albeit speculatively, the explosion with which the universe began. Assuming only that known laws of nature began to operate quickly, they thought they could give a coherent account of it. They relied on the 'times-ten' reckonings to find their way among the vast quantities involved.

However you looked at it, though, the Big Bang was a great melting pot, for human ideas about cosmic forces as well as for the matter and energy of the universe. For black-hole theorists, Einstein's gravity would imply that it started in a 'singularity', a geometrical point of no size, yet possessing enormous mass. In that condition, the known laws could not be valid. Yet, like an exploding black hole, it could create matter out of seeming nothingness and become, in an extremely brief moment, a seething and compact object packed with matter and energy, at unimaginable densities.

You might think it hard enough trying mentally to pack all the material of all the galaxies into an extremely small volume. Yet it was worse than that. The theory of the Big Bang indicated that vast quantities of anti-matter as well as matter existed in the first moment, exceeding by a billion times the matter which was to survive. Most of it cancelled out by mutual annihilation.

Two acute problems arose in trying to understand those events. One was a broken symmetry. There was slightly more matter than anti-matter, allowing the universe to take its eventual material form. The other was the perfect symmetry of electric charges, positive and negative, in the surviving universe, which found expression in an exact match in the numbers of electrons and protons.

Speculations about the anti-matter asymmetry included the idea that there was really as much anti-matter as matter in the universe, but it had just become separated into different lumps – different stars, different galaxies perhaps. But evidence for ongoing catastrophic encounters of matter and anti-matter was conspicuous only for its absence. Another suggestion was that a sister universe, or a group of universes, formed at the same time as ours and removed the excess anti-matter out of our sight.

But there might be a subtler and more satisfactory explanation for the broken symmetry. Abdus Salam emphasised that even protons could decay eventually into electrons. This was equivalent to the decay of a lump of radioactive material in which atomic nuclei broke up with overall predictability but not with precise regularity from moment to moment.

Salam argued that relatively slight statistical fluctuations in the decays of protons and anti-protons during the Big Bang could create the necessary small imbalance in favour of protons.

The strict equality of electric charges was then a consequence of the creation of matter and anti-matter in initially equal amounts. The Big Bang was not creating negatively charged electrons and positively charged protons as completely distinct operations: it did not have to get the numbers 'right' twice. Any imbalance in the numbers of protons and electrons would be automatically self-correcting, because a decaying proton inevitably produced an anti-electron which would annihilate the 'excess' electron.

Most of the deep physics of the Big Bang was concealed in the first moment. When the temperature was extremely high and the distances between particles were very short, all the cosmic forces may have looked alike, for the reasons I have mentioned. As the universe expanded and cooled, and the inter-particle distances became greater, the symmetry among the forces broke, much as cold water breaks its symmetry by forming a layer of ice on one surface. In less than a thousandth of a second from the start, the temperature would have dropped sufficiently for all the cosmic forces to operate 'normally'.

Steven Weinberg of Harvard University told how the various cosmic forces might have come into operation. Besides his role as author of the 1967 theory that linked the weak and electric forces, Weinberg pursued a strong interest in gravity and the theory of the Big Bang. And in speculating about what happened in the first frantic and creative moments of the existence of the universe, he drew together many of the ideas I have been discussing – not only in this chapter but throughout the book. As he put it:

'To build a universe may be a good deal easier than you think. I suspect that, at the beginning of the Big Bang, nature was very simple. Then, as the incredible temperature began to cool down, all the variety of forces and particles that we know about today began to appear.'

When the universe was so extremely hot that every little quark carried as much energy as a jet plane, gravity, the electric force and the strong and weak forces all had about the same strength and the same range. They were all in effect the same force. There was thus a high degree of symmetry among all the types of forces and all the types of particles that filled the universe.

But that lasted only for the briefest instant, Weinberg supposed. Almost at once, the perfect symmetry was lost, as the force of gravity 'froze

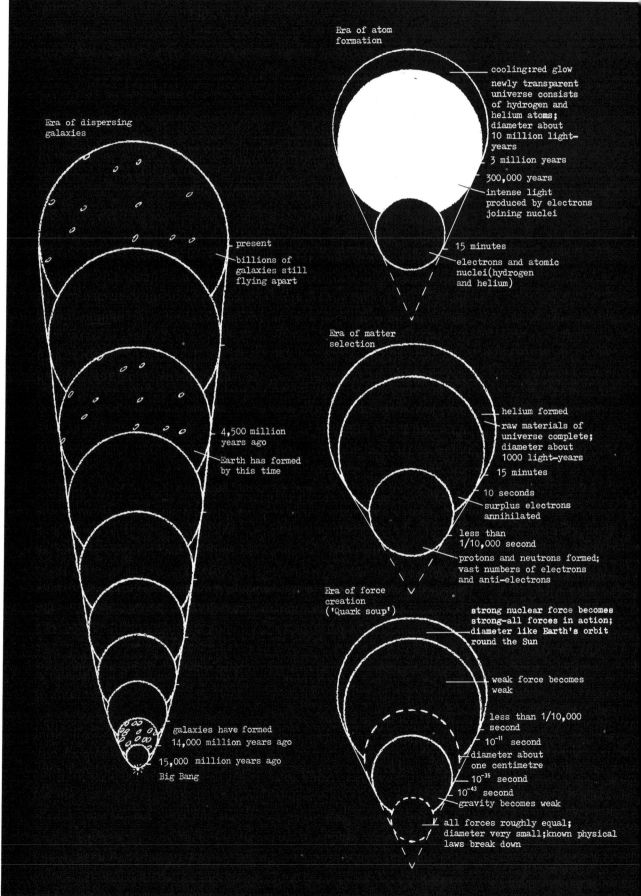

Era of atom
formation

cooling:red glow

newly transparent
universe consists
of hydrogen and
helium atoms;
diameter about
10 million light-
years

3 million years

300,000 years

intense light
produced by electrons
joining nuclei

15 minutes

electrons and atomic
nuclei(hydrogen
and helium)

Era of dispersing
galaxies

present

billions of
galaxies still
flying apart

4,500 million
years ago

Earth has formed
by this time

galaxies have formed
14,000 million years ago

15,000 million years ago
Big Bang

Era of matter
selection

helium formed

raw materials of
universe complete;
diameter about
1000 light-years

15 minutes

10 seconds

surplus electrons
annihilated

less than
1/10,000 second

protons and neutrons formed;
vast numbers of electrons
and anti-electrons

Era of force
creation
('Quark soup')

**strong nuclear force becomes
strong-all forces in action;
diameter like Earth's orbit
round the Sun**

weak force becomes
weak

less than 1/10,000
second

10^{-11} second

diameter about
one centimetre

10^{-35} second

10^{-43} second

gravity becomes weak

all forces roughly equal;
diameter very small;known physical
laws break down

out' and became weaker than the other forces. The electric force and the weak force remained in harmony for a while, both with the same strength and long range, until suddenly the weak force 'froze out' in its turn, and began to operate weakly over very short ranges.

As the universe continued to expand and cool, the pressure on the particles eased. When the temperature had dropped to just a few million million degrees the strong force began to make itself felt with all the quarks confined in their groups of three, like protons. Weinberg estimated that all this happened in the first ten-thousandth of a second after the start of the Big Bang.

No one could be sure whether this picture was historically accurate or not, but in a sense it did not matter very much. For Weinberg it illustrated the way that the unified 'gauge theories' of the cosmic forces were leading human beings to think about the present-day world.

'We have simply arrived too late in the history of the universe to see this primordial simplicity easily. ... But although the symmetries are hidden from us, we can sense that they are latent in nature, governing everything about us. That's the most exciting idea I know: that nature is much simpler than it looks. Nothing makes me more hopeful that our generation of human beings may actually hold the key to the universe in our hands – that perhaps in our lifetimes we may be able to tell why all of what we see in this immense universe of galaxies and particles is logically inevitable.'

The Big Bang and after

All that followed, after the first moment of the Big Bang, was just a fulfilment of the potentialities of the various particles and forces which began to operate as the pressure eased off. The universe went through several momentous eras, shaping its whole future in a very short time. In less than a thousandth of a second the universe cooled to a million million degrees and already the various cosmic forces were firmly in charge. The mutual annihilation of matter and anti-matter was already far advanced. It occurred much more rapidly for the heavy particles than for the electrons. It left the relatively small residue of proton-like particles that constituted the heavy matter of the universe and were to make hydrogen its chief raw material.

As the density of the universe diminished, those peculiar electrons without electricity, the neutrinos, withdrew into a world of their own, to march and countermarch ineffectually for ever through the widening universe. After 15 seconds the anti-electrons began to be annihilated. And in the first few minutes the strong nuclear force aided by the weak force changed a quarter of all the heavy matter of the universe from protons (hydrogen) into nuclei of helium.

The principal ingredient of the Big Bang was not matter but radiant energy – particles of light in the broad Maxwellian sense, from radio waves to gamma-rays. They kept the matter in a frenzied, electrified state. The universe was a fireball in which atoms could not form because their potential constituents were too agitated.

And the universe remained in that condition for hundreds of thousands of years – a period much greater than the split seconds with which we began the story, but still just a moment compared with the long subsequent history of the universe. The fireball was expanding all the time and cooling steadily, until the rays pervading it were no longer intense enough to keep matter so disrupted.

When everything was at about 5000 degrees, with the whole universe only as hot and bright as the present surface of the Sun, the matter of the universe disengaged itself, as it were, from the rays. It was not as emphatic a withdrawal from the action as the neutrinos had made, much earlier, but it allowed the matter to cool and form an ordinary gas. One electron would drop neatly into place around every proton to make an atom of hydrogen; two electrons teamed up with the heavier composites to make atoms of helium. That happened to all the ordinary matter in the universe. In the process, newly forming atoms gave off a great glow of visible light.

At that time, the universe was about 300,000 years old and it was about one two-thousandth of its present size: in other words, the material of the universe was still very tightly packed. During the subsequent expansion the light released by the formation of atoms became 'stretched' and reddened. The 5000-degree glow cooled to the 3-degree radio energy of the microwave background.

Gravity gathered the gas into great clouds and so formed galaxies. Conceivably the formation of galaxies required a certain 'lumpiness' in the Big Bang, or even a 'seeding' of galaxies with black holes. Whatever the precise process may have been, the galaxies came into being, and within them gravity shovelled the gas into denser knots to make the stars. The first stars were glowing in the first galaxies, perhaps one billion years after the Big Bang.

The great heat and pressure generated at the centre of each star by the infall of the gas set the strong nuclear force in action again converting hydrogen into helium, and releasing energy. But heavy stars burned quickly and exploded. Thanks largely to the work of the weak force, they scattered heavier elements – carbon, oxygen, iron, gold, uranium.

Gravity was still busy building stars, but the material it gathered was now enriched with heavier dust, from the explosions of the earlier stars. The Sun was one of a later generation of stars, and in its skirts gravity built planets from the dust, including the Earth. From radioactive elements excess energy, stored in the explosions of the ancestral stars, trickled into the rocks of the Earth's interior, heating them. That caused volcanoes to burst forth, filling the oceans with water and the atmosphere with gases. The heat also slowly shuffled the continents about the globe; when they collided the wreckage was a mountain range.

In the loose stuff – water, gas, dust – at the surface of the Earth, sunlight awoke the electric force. It made atoms combine and break apart, and out of this chemistry sprang life. For thousands of millions of years the nuclear reactor in the sky burned steadily. Life harnessed the sunlight as its source of energy and employed the electric force to combine atoms in ever more subtle ways: to breed, to fly, and eventually to power brains that wondered about their own existence.

Although some parts of that account were more vague or conjectural than others, it was remarkably comprehensive. Human beings might therefore flatter themselves that they understood rather well the workings of the universe and their own place in space and time, and sense a special communion between the material universe and the human mind.

A palpable key?

After thousands of years of speculation, three hundred years of modern science and the twentieth century's long revolution in physics, human knowledge of the deepest workings of the universe took a stride forward in the 1970s. Experimenters detected small objects inside the proton; a new cosmic force, the Starbreaker, revealed itself; and matter appeared carrying the new quality of charm. The value of these results was all the greater because they were predicted by the theoretical scheme which said the universe was built by quarks and electrons, and cosmic forces were carried by particles that shared their qualities. The new phase of physics which started with the discovery of strangeness twenty years earlier culminated in the discovery of charm.

But just how great a stride was it? Was Steven Weinberg justified in his high optimism about 'logical inevitability'? No one could tell. Theoretical ideas – not idle speculations, but reasonable extensions of the well-established scheme – were still running ahead of experimental evidence. The colour force, in particular, would not be easy to verify if the quarks were essentially coy. And before the astronomers had fully convinced themselves that they had found black holes among the stars and in the hearts of distant galaxies, Stephen Hawking was asking them to look for exploding black holes, for an effect that would help to unify gravity and the physics of particles.

The theories had certainly not arrived at their terminus. Benjamin Lee, the leading charmster at Fermilab, offered a specification for an ultimate theory that would encompass all known interactions – what I have called the cosmic forces.

'An ultimate theory would have only one dimensional constant, the Planck mass, and perhaps a cosmological constant. All other physical quantities should be computable. It must explain all natural phenomena – from galaxies to quarks and beyond.'

The 'Planck mass' was a computable mass of about ten millionths of a gram. That would be the smallest imaginable black hole – such that the diameter of the black hole was equal to the uncertainty about the location of the mass in space. In principle it might define the speed of light, the relationship of masses and charges in all particles, the strengths of the various forces, the limit to borrowed energy (Planck's Constant) . . . just about everything except perhaps the overall history and fate of the universe, which was why a 'cosmological constant' might be necessary.

All shades of opinion could be found, about how much closer the advances of the 1970s had brought the human species to possessing the key to the universe. Nor was it a simple matter of optimism or pessimism. If you expected rapid progress towards a dénouement, you might be gloomy about the future of physics itself and fearful that the great game could come to an end.

'Put your son into biology!' That summed up one point of view on the state of play. The enterprise of Newton was nearly finished. The final synthesis of gravity and particle physics was attainable and some unspecified young person, a latter-day Newton, was about to accomplish it. The accelerators and space telescopes would then confirm the theory's predictions so exactly that it would be hard to make a case for further expenditure on such devices. There would always be points of detail among the stars and the particles worth a little examination. But the liveliest and cleverest minds would turn to other fields to conquer – probably towards those inherently complex systems, like biology, which represented the next great challenge for science.

At the other extreme of opinion were the 'Chinese-boxers' who declared that nature itself would prevent the enterprise ever coming to an end. Inside every box of particles explored by experiment at any given time there would be another box, representing new wonders of the universe and new qualities of matter, to be revealed by the next generation of people and instruments. While theories of the universe might indeed get gradually closer to the 'real story' they could never attain it because there would always be missing information.

It was customary to recall how confident the Victorian scientists had been before nineteenth-century physics fell on the banana-skins of radioactivity, quantum theory and relativity. Whatever anyone could say in the twentieth century, about the universe or the fundamental constitution of its contents, might be rewritten in the twenty-first century.

Yet the scope of the possible rewriting diminished all the time. By the last quarter of the twentieth century, telescopes seemed to be spanning approximately the entire history of the universe. Machines for exploring the micro-universe were testing the behaviour of matter under stresses so great that nature itself surpassed them only in very peculiar conditions. Robust theories of the universe remained valid through a wide range of observational results or hypothetical assumptions. Any remaining mysteries belonged within a cosmos that was known in outline.

In the middle of the spectrum of opinion, you could say that the natural cut-off of the supposed 'ultimate theory', the Planck mass, was far beyond anything attainable with any instruments yet imaginable. The complete verification of the most wonderful theory would therefore be postponed to an indefinite future. But would anyone consider it worthwhile to go on indefinitely?

The answer to the question, 'Do we have the key to the universe?' would be eventually psychological. When youngsters were impressed enough by what their teachers told them about the physical universe and its contents to stop asking profound new questions, then, correctly or incorrectly, they would think they had the key. That might be regrettable, because for centuries physics had been the leader and model of human intellectual activity. Some said that you would simply have to go on asking searching questions, even if they were much harder to come by. For example: 'Can we infer any predecessor to our universe, despite the destruction of information in the Big Bang?' or 'Can we create forms of matter that nature itself never possessed?' The alternative might be mental indolence, like the 'Newtonian sleep' that settled on English physics and mathematics, when people had been overawed by Newton and yet neglected his concern about particles and the unknown forces between them.

Misgivings of that kind were for the future. In 1976 Lee could still conclude his speculations about the ultimate theory in saying, 'We are living in an exciting era. Let's all get back to work.' There was plenty to be done and the instrument-builders knew it better than anyone. Great space telescopes had to be engineered and put into orbit, which would sharpen the views of the universe given by visible light and X-rays. Even while CERN at Geneva was completing a giant accelerator to match Fermilab's, Fermilab itself was doubling its accelerator's energy by equipping the main ring with superconducting magnets. Work had also begun on the next generation of machines, starting with electron annihilators much more powerful than SPEAR: PETRA at Hamburg and PEP at Stanford. Particular schemes of the machine-builders, notably at Fermilab and Brookhaven, were directed to the task of discovering the very heavy W particles predicted by the theory of the weak force: finding them would be the best possible confirmation of the gauge theories.

Already machine-builders from different laboratories around the world were beginning seriously to discuss a World Machine – an accelerator so big that even the United States would not afford it alone. It might give protons 10 million units of mass-energy, compared with up to 500,000 at Fermilab. Others, recalling the success of SPEAR, preferred the idea of annihilating electrons with 200,000 units of combined mass-energy. Whatever its character the World Machine would be an international collaborative venture, like CERN in Europe, but now extended to all nations willing to contribute to an engineering budget of about a billion dollars. 1990 seemed a possible target date for this expedition across another new ocean of energy. It was canvassed as a scheme that might fire the imagination of those concerned with international good-will, as well as the experts anxious to know if there were any new quarks to be found.

The big pay-off from fundamental physics would not, in fact, be a cosmic formula for the school-books of the future. It would be, in part, an assertion of mankind's trust in tested knowledge, rather than superstition or ideology.

Beyond that it was a gift to human self-esteem. Physicists glimpsed amazing possibilities for the troubled species if only it would raise its eyes from the furrow and the ledger. As Freeman Dyson of Princeton put it:

'Life may succeed against all the odds in moulding the universe to its own purposes. And the design of the inanimate universe may not be as detached from the potentialities of life and intelligence as scientists of the twentieth century have tended to suppose.'

The Space Shuttle, the
...-usable American
...uncher, brings in an
...a of easier spaceflight,
...lowing large
...struments to be put
...to orbit for studying
...e universe. Will large
...mma-ray detectors
...nd Stephen Hawking's
...ploding black holes?

Picture Credits

Acknowledgement is due to the following for permission to reproduce illustrations, (references are to page numbers).

2, David Rooks, University College London; 6, CERN; 8, Novosti; 9, Fermilab; 10, Fermilab; 13 top, Stanford Linear Accelerator Center, bottom left, CERN, bottom right, National Accelerator Laboratory; 20, Radio Times Hulton Picture Library; 24, Camera Press; 25, Cambridge University Press; 30, CERN; 34, Hale Observatories; 40, CERN; 42 and 44, Gargamelle Collaboration; 47, Argonne National Laboratory; 48, CERN; 56, Alec Nisbett; 57, Ric Gemmel for *New Scientist*; 58, Alec Nisbett; 64, CERN; 68, Karsh/Camera Press; 69, California Institute of Technology; 72, Brookhaven National Laboratory; 79, Stanford Linear Accelerator Center; 88, CERN; 91, Alec Nisbett; 96, Lawrence Berkeley Laboratory; 99, Alec Nisbett; 103, Brookhaven National Laboratory; 104, CERN; 106, Lawrence Berkeley Laboratory; 108, Stanford University; 110, Stanford University; 116, CERN; 118, Alec Nisbett; 121, Alec Nisbett; 123, Fermilab; 126, Fermilab; 134, Netherlands Foundation for Radio Astronomy; 147 top, Appleton Laboratory, bottom, Fotoarchief Sterrewacht Leiden; 150, Cavendish Laboratory; 152 top, *Nature*, bottom Sterrewacht Leiden; 155 top left, National Radio Astronomy Observatory, top right, Hale Observatories, bottom left, California Institute of Technology, bottom right, Max-Planck Institute for Radio Astronomy; 166, Alec Boksenberg; 179 top left and bottom right, A. F. Kersting, bottom left and top right, Robert Harding; 191, Associated Press.

Diagrams by Diagram.

Further Reading

Most of the story I have related is new and I know of no other books that yet cover quite the same ground. Detailed accounts of parts of it are to be found in scientific magazines, *Scientific American* and *New Scientist* for example, and in learned journals such as *Physical Review Letters, Physics Letters, Nature* and *Reviews of Modern Physics,* but almost always using a more technical vocabulary than I have permitted myself.

The following four books are up to date on important sectors of the story and are written in fairly plain language.

Adventures in Experimental Physics, Epsilon Volume edited by Bogdan Maglich (World Science Education, Princeton, 1976). It contains personal accounts of the discovery of the 'gipsy' (J/psi) particle by Ting, Goldhaber and Richter, as well as other relevant articles.

The First Three Minutes by Steven Weinberg (Basic Books, New York, 1977). A detailed description of the Big Bang by a leading theorist active in both particle physics and cosmology.

Cosmology Now edited by Laurie John (BBC, London, 1973). A series of radio talks by leading astro-theorists, including Bondi, Sciama, Penrose and Rees.

Astronomy and Cosmology by Fred Hoyle (W. H. Freeman, San Francisco, 1975). Although a textbook, it is as readable as you would expect from one of the best of scientist-writers, who has also been a leader, albeit an heretical one, in recent cosmological and astrophysical theory.

The books in the next list are older in content, but nevertheless give important information about past discoveries and abiding attitudes which made the recent advances possible.

The Character of Physical Law by Richard Feynman (BBC, London, 1965; paperback ed., MIT Press, 1967). A series of television lectures on the nature of physics by an outstanding theorist who is also generally acknowledged by his colleagues to be the best expositor of their subject.

Physics and Man by Tor Ragnar Gerholm (Bedminster Press, Totowa, 1967). Another thoughtful account of the character of modern physics by a noted Swedish experimentalist concerned about the cultural connotations of his subject.

The Nuclear Apple by Paul Matthews (Chatto and Windus, London, 1971). An account of high-energy physics for the layman, with special emphasis on the fall of parity, the development of the quark theory and discovery of the omega particle.

Einstein by Jeremy Bernstein (Viking Press, New York, 1973). A biography by a physicist who is also an excellent writer able to convey the scientific meaning of Einstein's work.

Violent Universe by Nigel Calder (BBC, London, 1969 and 1973). A report on the heady phase of astronomy which brought in the pulsars, quasars and microwave radio background and irrevocably altered the human perception of the universe.

Index

Entries in italics refer to mentions in captions only